Class

Inverness to Dover Western Docks, 1985–86

IAN McLEAN

BRITAIN'S RAILWAYS SERIES, VOLUME 22

Front cover: Saturday 20 September 1986. Eastfield's 47617 *University of Stirling* is far from home on 1C35 13.34 Norwich–Liverpool Street, which it will work to Ipswich for a Class 86 forward.

Back cover: Thursday 25 July 1985. The elusive 47595 *Confederation of British Industry*; two weeks solid in Scotland and neither hide nor hair of 47595 had been seen. Once back across the border on our way home, there was the Eastfield loco at the head of 1E20 10.40 Carlisle–Leeds. It had been in England all along, working 1E86 07.20 Blackpool–Cambridge on 9 July, before failing at March with binding brakes on 1M12 14.10 Cambridge–Blackpool the next day. After repairs at Stratford, it worked 1S85 07.17 Harwich–Glasgow to Preston on the 11th and returned east on 1E87 11.15 Glasgow–Harwich the next day, before working 1S85 again on 13 July.

Title page: Saturday 26 October 1985. 47550 *University of Dundee* clatters over Haymarket West Junction with 1B07 11.18 Carstairs–Edinburgh, a portion detached from 1S45 07.50 Manchester Victoria–Glasgow, which also had a portion from Liverpool that had joined at Preston.

Contents page: Wednesday 2 April 1986. Eastfield's 47578 *The Royal Society of Edinburgh* had decided to take its summer holiday early this year. It had worked 1E75 14.14 Exeter–Leeds the previous day, having worked to the West Country on the Sunday on 1V84 11.15 Liverpool–Penzance to Exeter. Earlier this morning, it had worked 1O27 06.45 York–Portsmouth Harbour to Birmingham and then stuck to diagram by taking the 1O74 09.58 Manchester–Brighton forward from Coventry, on which it is seen at Oxford. It later completed its day by working 1M41 18.48 Brighton–Derby throughout. The next day saw it back at Brighton to work 1M07 12.55 Manchester to Brighton, suggesting that it must have worked 1O62 06.02 Derby–Brighton throughout. Its holiday continued with visits to Poole on the 4th and 6th.

Published by Key Books
An imprint of Key Publishing Ltd
PO Box 100, Stamford
Lincs, PE19 1XQ

www.keypublishing.com

ISBN 978 1 913870 43 0

Typeset by SJmagic DESIGN SERVICES, India.

Contents

Introduction

When I started 'bashing' in 1982, 507 of the 512 Class 47s built were still in service. On 14 August 1985, the final day covered in book one, I scored 47229, which was my 459th, so there were still 48 to go. However, the locos were getting older and those involved in heavy collisions would not necessarily be repaired as they would have been in the past. The race was on!

I was right to be concerned as, on 19 January 1986, 47111 was destroyed when it was hit by a runaway DMU on a football special at Preston. Repairs were costed at £60,000 but were not authorised. There would be no reprieve for 47111, which was officially withdrawn on 17 March 1986, and I had missed another one.

Barely a month later, on 27 April 1986, 47282 was involved in a fatal collision at Litchfield Tunnel, near Micheldever, when it hit the back of a ballast train while running light to Eastleigh. The force of the impact crushed the cab and derailed five wagons. That was another one lost.

The start of 1986 would also see the withdrawal of the first Class 47/4s, when three 'Generators' (47401–420) were condemned to keep the other 17 of these non-standard locos going. This was the thin end of the wedge, as the start of 1987 would also see serviceable '47/4s' withdrawn when they became due for overhaul, with 47429 and 47529 being sacrificed to provide a spares pool for the upcoming Cost Effective Maintenance (CEM) programme.

We were also entering a period of great aesthetic change, as the Corporate Blue era gave way to a plethora of new colour schemes. Several Scottish Class 47s had already received depot repaints into Large Logo livery, and the Class 47/7s had been given ScotRail livery at Crewe Works, or at depot, as they were rebranded as a matter of urgency. 47422, 47438 and 47503 were the first to get Large Logo after overhaul at Crewe in late 1985 and emerged with hand-painted numbers, as no transfers had been made at the time. 47461 was outshopped from Crewe in InterCity livery in August 1985; the same month saw 47050 lose its boiler at overhaul at Crewe and reappear in the new Railfreight livery. The spread of this scheme took slightly longer than the authorities had hoped for as Crewe took the instruction to repaint non-boilered '47/0s' a bit too literally and only adorned those in the 470xx number series with the new livery!

Another tranche of Electric Train Heat (ETH) conversions, starting with 47629, began to emerge from Crewe, in Large Logo livery, from October 1985 and these would sound a death knell for steam heat. Christmas 1985 was a watershed, as the numerous relief trains that had, in the past, been steam-heated were now electrically heated. By 1986, steam heat was becoming very rare in England and regular trips to Scotland became the norm in order to enjoy this atmospheric, if sometimes unreliable, form of train heating. One long-distance steam-heated diagram remained between Edinburgh and Inverness, but ETH even started to appear on the Edinburgh–Dundee circuit. Steam heat hung on, mainly on football specials, until May 1987, when 47117 steamed for the final time.

July 1986 saw the emergence from Crewe Works of 47650, the first of the conversions with long-range fuel tanks to eliminate the need for loco changes on King's Cross–Aberdeen overnight trains. These extended range machines also allowed for the loco change at Plymouth on the Newcastle–Penzance service to be dispensed with.

Not too far into the future, sectorisation would also start to kick-in from April 1987, with many Class 47/4s no longer being regarded as passenger locos as they passed to the Parcels, Departmental and even Railfreight sectors. Thus, some '47/4s' would now be rare on passenger trains, something that could not have been envisaged a couple of years earlier.

On a personal level, 6 December 1985 proved to be a fateful day; 47422 took a group of us from the East Midlands to Harwich on the 'European' for a weekend in Holland. On the trip I met a bloke called Vince for the first time. Vince worked on the railway, and he suggested that I should apply for a summer job on the railway and told me who to apply to. I took his advice and found myself working in the Travel Centre at Derby in the summer of 1986. As well as the very welcome money, the job also brought with it Privileged Travel (Priv) after a qualifying period of one month. This would make bashing much easier with Rover tickets now being quarter price! Part one of my railway career would only last for the summer months, though, before going to university.

The primary objective was, of course, to try and get all the Class 47s for haulage, but other games took place too, such as trying to get '47s' to different places. New places visited with the class in this book include Matlock, London Euston, Dover Western Docks and the delightful Cardenden.

The tools of the trade

The 'Moves Book', which kept a record of trains travelled on, between which points and the locos, the national timetable, Loco-Hauled Travel, the Motive Power Pocket Book and 1H85, which listed all loco-hauled trains and the type of stock they were formed of.

Chapter 1

Summer 1985 continued

Tuesday 23 July 1985. A rediscovered negative of a photograph that haunted me for years, as it seemed that I would never have a Class 47 northwards from Inverness. After having 47550 *University of Edinburgh* on the internal overnight from Edinburgh, 47120 *RAF Kinloss* was in the Far North platform with the empty coaching stock (ECS) to form 2H68 07.55 Invergordon–Inverness. It was generally acknowledged that a friendly word with the guard would see permission granted to ride on the empties to Invergordon, but with 47005 on the first one 'over the top' to Aberdeen and the extra mileage offered, we were tempted eastwards instead. Note the steam heat in operation, even in July! The Far North situation was finally rectified in 2006 when a group of us did 'The Easter Highlander' to Kyle of Lochalsh and Thurso.

Friday 23 August 1985. With a scar on its flank where the *Orion* nameplate used to be, 47083 pauses at Bristol Parkway on 1V52 10.20 Edinburgh–Plymouth relief, having earlier worked to the second city on the 1Z09 10.41 Exeter–Edinburgh relief. After an HST back to Birmingham, the day was rounded off with 47333 on 1V16 17.25 Manchester–Paddington to Banbury for Thornaby's 47360 home on 1M41 18.48 Brighton–Derby. 47083 arrived at Crewe Works at 10.50 on 10 October for Intermediate overhaul and ETH fitment. It would be released on 19 December as 47633 (page 28) and subsequently transferred to Inverness!

Saturday 24 August 1985. Bowled by 47333, if such a thing is possible. The day started with 47360 and 47564 in tandem on 1O62 06.00 Derby–Brighton to New Street, for 47533 forward in order to intercept 1M88 06.33 Poole–Manchester at Oxford, expecting it to be a required 47068, only for 47333 to roll in. It was still a good day, having 47357 to Basingstoke on 1O13 07.14 Leeds–Poole, for the required 47377 back to Birmingham on 1Z48 11.30 Poole–Wolverhampton relief. 47315 was also scored later in the day.

Friday 13 September 1985. A train that always had to be watched was the Fridays-only 1E22 12.46 Portsmouth Harbour–Leeds. Although not booked to run round at Birmingham, on this occasion it had with Crewe's 47341 working the train throughout. The loco later worked 1O29 22.32 Leeds–Portsmouth Harbour back to Birmingham, where it was remarkably replaced by another '47/3' in the form of 47307.

Saturday 14 September 1985. A couple of drivers share a story as they await the arrival of 1O27 06.45 York–Portsmouth Harbour, which 47249 will work forward. Although booked for a Gateshead '47' into Birmingham, the train struggled in behind 31417, which would give Control a problem, as the inward loco was booked to run light to Coventry to work 1O74 09.58 Manchester–Brighton to the south coast. In the end, 47611 was found to work 1O74 forward. 47249 was taken to Leamington, after which my next three '47s' would all be scores.

Immingham's 47212, which, like most Class 47s from this depot, was very rare on passenger trains, became my 469th class member back to New Street on 1M03 08.05 Portsmouth Harbour–Manchester. The engine for 1M03 was booked to come off an overnight Freightliner to Southampton and, as such, it was generally a no-heat locomotive. From Birmingham, it was a case of needs must and an HST to Chesterfield to intercept Tinsley's 47372 on the last leg of the 1E02 10.43 Skegness–Sheffield. It may only have been 12¼ miles, but it was a good 12¼, with 47372 ripped open up the climb to Bradway Tunnel, followed by a fast run on the 90mph descent to Sheffield. The day was rounded off by going to Doncaster for Bristol's 47159 (471st) on 1L19 13.30 Skegness–Leeds.

Saturday 21 September 1985. It was rare to see a Gateshead loco looking this shiny! As D1550, the loco had been the first Class 47 to be constructed at Crewe Works and, as 47435, it had just enjoyed what would be its final Intermediate overhaul at its birthplace and, remarkably, it had emerged on 5 August with its train heating boiler intact. On the autumnal equinox, it is seen at Wolverhampton waiting to work 1J24 09.35 Euston–Aberystwyth, forward to Shrewsbury. Typically for a Gateshead loco, it had been hard at work since its release from works, having visited Brighton, Portsmouth, Poole, Paignton, Penzance, Harwich, Preston, Liverpool, and Newcastle, amongst other places.

When things go wrong. The 1E14 10.05 Portsmouth Harbour–Leeds had been 47377 into Birmingham for 47348 forward via Leicester and Nottingham. By Leicester, 1M67 10.42 Yarmouth–Birmingham still had not gone past due to the failure of the Class 31s on the train. Haymarket's 47013 had replaced the Type 2s at March and the idea was to have it back to New Street. However, 1M67 was running so late that the unusual step was taken to terminate it at Nuneaton, hence all the open doors and bodies on the platform.

47366 arrived from Saltley to work 1E80 16.20 Birmingham–Norwich starting from Nuneaton with the stock off 1M67, but with no Class 47 to return on, the best option was to wait at Nuneaton for 1M68 12.35 Yarmouth–Birmingham, which at least was on time, with Crewe's 47135, with its faded grey roof from its time at Stratford, at its head. This was the infamous night when 47537 failed at Soho on 1S19 16.42 Penzance–Glasgow, just two miles after it had been put onto the train. To the horror of those on board, the train was rescued by 31420, which struggled to Bescot where another nasty surprise lay in store as 45016 was put on the front. A nightmare of a night.

Sunday 22 September 1985. The next day was not much better either. 47056 and 47447 were on the Nuneaton drags and I ended up stuck at New Street for about five hours with not a lot happening. Eventually, out of sheer desperation, Inverness's 47550 *University of Dundee* was taken to Stafford via Cannock on 1S61 07.34 Poole–Glasgow, just for a change of scenery. A couple of well-known bashers of the time, Oakham (who came from Oakham) and Stafford (who didn't hail from Stafford), are on the platform.

The move to Stafford had been speculative to see if there was something no-heat on the 1G06 13.10 Manchester–Birmingham, but it was 47515. 47430 was on 1E92 14.53 Bristol–Leeds to Derby, where continuing the Scottish theme, 47017 was employed on a ballast job at Derby, heading for Chaddesden Sidings. The Haymarket loco had been released from Intermediate overhaul at Crewe Works on July 24 but failed after working 2J02 12.30 Dundee–Edinburgh on 3 August and was returned to Crewe for rectification. Released again in mid-August, it was showing a reluctance to return home.

Saturday 28 September 1985. The final summer Saturday of 1985 and, once again, the Eastern Region were causing problems for the Midland by sending out 45143 on 1O27 06.45 York–Portsmouth Harbour. The loco off 1O27 was booked to work the 1O74 09.58 Manchester–Brighton forward from Coventry but, in a clever piece of retribution, Stratford's 47570 was pinched to work 1O74 instead, having earlier worked 1Z27 03.30 North Walsham–Aberystwyth charter to Shrewsbury Abbey Foregate Street. The Eastern were no doubt surprised to get 47598 rather than 47570 back that night. The secondman checks that all is well as 47203 departs Coventry on 1O27.

Disappointingly, the day was a bit of a damp squib with no winners out. It had not started well with 1O62 06.02 Derby–Brighton worked out of Derby by 47622 and 45127 in tandem. These were replaced by 47626 *Atlas* at Birmingham for the run to the south coast. 47421, which had been named *The Brontës of Haworth* on 8 August, was on 1M10 06.58 Paddington–Llandudno, 47079 *GJ Churchward* had worked 1E66 08.10 Birmingham–Yarmouth to Norwich and the rest of the diagram, with 47590 *Thomas Telford* on 1E72 10.20 Birmingham–Yarmouth, in place of the booked pair of Class 31s. The 1M03 08.05 Portsmouth Harbour–Manchester, which was normally the loco off an overnight Freightliner service, was 47202, which had worked 1O29 22.32 Leeds–Portsmouth Harbour. 47032 worked 1E14 10.25 Portsmouth Harbour–Leeds to Birmingham in its place, with 47338 working 1E14 forward. I ended up heading home on 1E63 10.40 Poole–Newcastle, which was 47522 into Birmingham for 47524 forward. At Derby, Eastfield's 47005 rolled in on 1E64 09.55 Penzance–Leeds, in place of the booked Class 45, so a blast to Rotherham finished my summer off. I ended the summer with 472 Class 47s in the book, with potentially another 25 to get.

What we did not know at the time was that 1985 would be the last summer of widespread non-ETH locos on summer Saturday trains. There would only be three booked non-boiler Class 47s diagrams in the West Midlands in 1986, with another one between Sheffield and Skegness and another between Preston and Blackpool North.

So, we moved into the dark, cold winter months. Bashing in the winter months was a much less stressful affair, the fear of missing a required loco was greatly reduced and there was better opportunity to accumulate mileage. Disappointingly, the £3.75 fares from Birmingham to the south coast did not happen again, but the Fabulous February Flings more than compensated.

Chapter 2

Winter 1985–86

Saturday 26 October 1985. 47710 *Sir Walter Scott* runs into Haymarket with 1O22 11.30 Glasgow–Edinburgh, which was timed for 42 minutes for the 46 miles from Glasgow, including a stop at Falkirk High. 47704 *Dunedin*, 47708 *Waverley* and 47716 *The Duke of Edinburgh's Award* were the other three locos on the Edinburgh–Glasgow route on this day.

Before and After 1

Saturday 26 October 1985. Gateshead's 47482 has arrived in Edinburgh with 1B32 12.00 from Aberdeen and is in the process of shunting three vans off the front of the train. One of the few still sporting a 'domino' headcode by this time, the loco would be stopped for Intermediate overhaul at Crewe Works on 5 December.

Saturday 1 March 1986. 47482 re-entered traffic on 21 February and has been transformed since the photo taken on 26 October 1985. The immaculate loco is seen arriving at Edinburgh about an hour late on 1S77 23.35 sleepers only from King's Cross. As a Gateshead loco, it would not be expected to stay in this condition for long.

Before and After 2

Sunday 27 October 1985. Crewe-based 47454 stands in Platform 1 at Glasgow Central at the head of 1M19 14.05 to London Euston, which it will work to Carlisle via Dumfries due to booked engineering work. 47454 was stopped on 9 March 1986 but did not actually arrive at Crewe Works for overhaul until 4 April.

Saturday 14 June 1986. Fresh out of the paintshop, 47454 had only been released from Crewe Works on the evening of 11 June following overhaul. Seen here at Chester at the head of 1D64 15.10 Euston–Holyhead, the train has a Mk.3 sleeper for use on 1A04 01.15 Holyhead–Euston as the first vehicle. Had I stayed here, I could have had 47148 on 1K34 17.40 Llandudno–Crewe, but I continued west 'blind' for 1K36 18.00 Holyhead–Crewe, which had 47551 at its head.

Sunday 27 October 1985. With the loco unusually on the west end, 47704 *Dunedin* awaits departure time with 1O83 15.30 Edinburgh–Glasgow, which was booked to Glasgow Central via Shotts on this date due to engineering work.

Saturday 9 November 1985. Gateshead's 47424 sits in No 1 through siding at Birmingham New Street awaiting the arrival of 1O25 09.45 Glasgow–Poole, which it will work forward to Reading. 47424 and 47427 were unusual in that they lost their boilers when they were converted to ETH, while all of those around them retained theirs. This was because they were LMR locos at the time, while the others were ER locos, and the boilers were kept largely for steam-heated overnight work. Note the enthusiast on the platform who has brought along a camping seat to offer a modicum of comfort while he waits for events to unfurl.

Saturday 16 November 1985. 47551 pauses at Chesterfield on 1O19 08.00 Newcastle–Poole on a wet, horrible day. The Eastern Region was always reluctant to send out a Class 47 on 1O19 as, more often than not, it would not get it back on 1E63 09.40 Poole–Newcastle. On this date, 47441 went forward from Birmingham on 1O19 but, sure enough, the Midland kept 47551 and sent 45112 back in its place. 47551 had transferred from Stratford to Gateshead six months previously and, despite its grey roof, it was now looking much more like a Gateshead loco than a Stratford one.

Saturday 30 November 1985. Two weeks later and the Eastern Region had once again sent out a Class 47 on 1O19. However, on this occasion they got it back! Gateshead's 47497 waits time at Derby on 1E63 after a typically fast run. 47567 had replaced it at Birmingham on 1O19 and 47497 had, in turn, replaced 47488 on 1E63. After 47497 to Sheffield, 47565 was had back to Chesterfield on 'The European', the 1E87 11.15 Glasgow–Harwich Parkeston Quay. Something must have failed in East Anglia as 47580 *County of Essex* appeared on 1M12 14.10 Cambridge–Blackpool North and was taken to Manchester Victoria for 47481 back on 1M87 15.15 Glasgow–Nottingham. You could see why Liverpool Street Control did not like losing its '47s'; by the night of Friday 13 December, 47580 was in the north-east of Scotland working 1E43 20.35 Aberdeen–King's Cross!

Saturday 21 December 1985. Although it was part of Gateshead 16 diagram, 1E54 06.50 Paddington–Hull had a nasty habit of producing a Class 45 from New Street, as the loco started the day on 1O62 06.02 Derby–Brighton. All looked set fair as 47433 worked 1E54 north and, at Sheffield, 47556 was found to be on Gateshead 13 diagram to work the portion off 1E54 forward as 1L17 11.35 Sheffield–Leeds. Letting 47433 go, I opted for 47556 to Leeds, which was a mistake! Upon arrival at Rotherham, 47556 shut down and refused to restart. Then ensued a very long wait hoping that something good might arrive from Tinsley; it didn't. A very disappointing 31401 arrived and I collected my things from the front coach. There was to be no escape though, as I finally left Rotherham at 13.33 behind 31455 on 1M31 12.08 Hull–Manchester.

Monday 23 December 1985. The Christmas reliefs of 1985 had a very different feel to those of previous years. In the past, they had been a festival of steam heat, but this year ETH dominated. On the Friday, I had 47549 on 1Z37 13.25 Reading–Crewe, on the Saturday it was 47613 *North Star* on 1E50 11.00 Plymouth–York, and today it was the turn of 47489 on 1V32 10.30 York–Plymouth, seen here at New Street. 47489 had been released from Intermediate overhaul at Crewe on 13 November, but was already looking work stained.

Friday 27 December 1985. Unusually, Liverpool Street Control had sent out one of its own on 1S85 07.17 Harwich Parkeston Quay–Glasgow. 47591 has run round the train at Sheffield in readiness for the run through the Hope Valley. In a busy scene, a Class 31 runs into No 3 siding on the Cleethorpes–Manchester empty NPCCS, while a DMU sits on the through road. In another illustration of why the Great Eastern did not like sending its locos out on 1S85, rather than returning on 1E87 the following day, it was on the 1M72 16.05 Leeds–Carlisle instead. The photo was taken from 1E51 07.52 Exeter–York relief, hauled by 47575 *City of Hereford*.

At York, 47422 was viewed on an Edinburgh–King's Cross relief. 47422 had emerged from Intermediate overhaul, during which it lost its boiler, at Crewe Works on 18 October with hand-painted running numbers, as none of the new-sized ones were in stock. It had not lasted long in traffic, failing on 25 October, and returning to Crewe for rectification, which saw it side-lined until 19 November. It had been a solid performer since then; taking 'The European' into Harwich on 18 December and working a relief into Inverness on Christmas Eve. Note the Class 03 and 08 in the former Motorail dock, which is now part of a car park.

The 1V21 14.25 York–Plymouth relief also produced an ETH Class 47 in the form of 47538, the erstwhile *Python*, standing in Platform 15, which became Platform 10 when the station was remodelled. The loco was named without ceremony at Old Oak Common on 31 March 1966, but the plates had gone by September 1973, sadly never to be refitted, although replacement plates are believed to have been cast.

Saturday 28 December 1985. Swindon Control had obviously thought that 1E92 11.00 Cardiff–York relief was a good way to repatriate 31460, only for the task to prove too much for the Type 2, which capitulated at Bromsgrove at the prospect of the climb to Blackwell. Assistance came in the form of 47339, which worked through to York, with the Class 31 heating the train. Having proved to be a late Christmas present at Birmingham, 47339 opens up on departure from Sheffield. It was quite unusual by this time for a Class 47 not to have a headlight; 47339 would be the penultimate one fitted, with 47008 being the last.

Sunday 5 January 1986. Thornaby's 47362 runs light south through the Goods Lines of a snowy Derby station. 47362 was the sixth Class 47 to be painted into Railfreight colours at Crewe Works during its November 1985 Intermediate overhaul. 47050 had been the first, followed by 47095, 47378, 47237 and 47374. More would have preceded it, but Crewe Works misunderstood the livery orders and only locos in the 470xx series were being thus painted at first.

Sunday mornings always meant 'drags', electrically hauled trains being towed by a diesel when diverted over non-electrified lines or under de-energised wires. 47234 and 47602 *Glorious Devon* were employed on Nuneaton drags this day, while further north 47409 *David Lloyd George* and 47561 were on drags between Manchester and Stafford. 47561 has just been coupled to 86225 *Hardwicke* at Stafford on 1H32 11.50 Euston–Manchester. There was further emergency dragging elsewhere as 47196 worked 1A29 16.00 Liverpool–Euston to Crewe via Warrington, due to an electrical isolation, and returned on 1F41 16.55 Birmingham–Liverpool. I made my way back behind 47439 on 1G06 13.10 Manchester–Birmingham.

Saturday 11 January 1986. Sheaf House dominates the skyline as Stratford's 47566 departs Sheffield on 1E54 06.50 Paddington–Hull, after the train has been split in half. As can be seen, Gateshead 13 diagram has failed to produce a Class 47 yet again and a Class 31/4 waits to work 1L17 to Leeds. 47566 was booked to work 1J01 13.24 Hull–Sheffield and 1L18 18.01 Sheffield–Leeds, but York Control took the opportunity to swap things over and send it home on 1E87 11.15 Glasgow–Harwich Parkeston Quay from Sheffield, rather than the incoming 47453 running round.

Saturday 25 January 1986. The careers of 47542 and 47544 seemed to be entwined, as it always seemed to be a case of if you saw one, the other would not be far away. They both moved from Gateshead to Stratford on 2 October 1983 and both would go back to Gateshead on 16 March 1986. 47542, which had received the trademark grey roof by September 1984, waits to drop onto the stock to form 1H14 10.35 Liverpool Street–King's Lynn.

Speak of the devil! Lurking over the other side of the station on a van train was 47544. It was quicker to get a grey roof and had it by March 1984, only to lose it at its April 1985 overhaul, but it was reapplied by October 1985. The careers of the pair diverged drastically on 15 May 1988 when 47542 was transferred to Laira, while 47544 moved to Eastfield. Both still retained their 'Stratford' roofs at the time.

Liverpool Street was a building site at the time, with the platform that was normally used by the Norwich services out of commission and occupied by an engineers' train with 47291 at its head. 47291 had gained a grey roof in a previous stay at Stratford but did not get one during its September 1985 to May 1989 residency. 47591, on the other hand, had gained one by September 1983 after its conversion from 47265 in July 1983. 47591 would later work 1H26 15.35 Liverpool Street to Cambridge.

Saturday 1 February 1986. And so, the adventure began. During February, BR was doing an offer that saw fares capped at £12 with a Young Persons Railcard. The day started with 47434 on 1G14 07.17 Derby–Birmingham for 47616 *The Red Dragon* on 1M10 06.03 Paddington–Liverpool. The driver of 47468 walks towards his steed prior to the departure of 1P11 11.15 Liverpool–Preston, after which it will work 1E87 11.15 Glasgow–Harwich Parkeston Quay to its destination.

After, 87029 *Earl Marischal* was on the 'Royal Scot' for the run over the Fells. 47534 arrived in Carlisle on 1M06 11.00 Stranraer Harbour–Euston, and 47404 *Hadrian* was provided to work the corresponding 1S74 10.15 Euston–Stranraer Harbour, seen here with a youthful-looking driver waiting for the tip to leave Kilmarnock. The 16.45 DMU would take me on to Glasgow.

Arriving in Glasgow Central at 17.35, a stroll across town to Queen Street revealed 47716 *The Duke of Edinburgh's Award* at the head of the 18.00 to Edinburgh and 47464 on 1L39 18.05 to Perth. After its trip to the 'Fair City', 47464 is seen back at Queen Street off 1T40 19.34 Perth–Glasgow. Sadly, 47464 was in its last year of service, being damaged beyond economical repair on 23 September when it was hit by 37416 near Elgin.

Sunday 2 February 1986. The overnight route back south on 1V62 23.35 Glasgow–Bristol was diesel hauled throughout due to engineering work, this starting off with 47503 via Beattock to Carlisle, where it was replaced by 47413 to Preston via the S&C. A reversal then saw 47534 work to Birmingham, but it was piloted by 47344 from Stafford over the Cannock route. At New Street, there was the option to head back north with 47446 on 1P06 09.27 to Blackpool, but heading south was the better option.

47488 was taken south on 1O62 09.05 Wolverhampton–Gatwick Airport instead. At Gatwick, it was announced that the train was to continue onto Three Bridges, so naturally we stayed on. This then gave us a problem as the train was ECS back to Gatwick, and the guard was initially unwilling to let us back on. Fortunately, after a bit of persuading, he acquiesced and 47488 is seen at Three Bridges, on the not quite ECS to Gatwick.

Saturday 8 February 1986. The day started with 47440 on 1S49 07.25 Nottingham–Glasgow to Preston, where it was replaced by an AC electric for the run over the Fells to Carlisle. The journey north continued with 47533 on 1S37 12.40 Carlisle–Glasgow via the GSW route. Another stroll across Glasgow found 47704 *Dunedin* on 1A63 15.25 Glasgow–Dyce and 47714 *Grampian Region* on 1O38 15.30 Glasgow–Edinburgh, with the latter taken to Haymarket in the search of some steam heat on 'the circuit'.

Dropping straight in, 47049 was on 2J27 16.15 Edinburgh-Dundee and was taken to Markinch for 47003 on 2J44 16.30 Dundee–Edinburgh back to 'Auld Reekie'. A glimpse into the future then saw recently converted 47637 on 2J31 18.15 Edinburgh–Dundee, which was taken to Inverkeithing for 47018 on 2J46 17.30 Dundee–Edinburgh. 47560 *Tamar* on 2J29 17.15 Edinburgh–Dundee completed the domination of 'the circuit', but two ETH locos were a sign of things to come. A Class 27 did appear later when 27017 worked 2J33 20.15 Edinburgh–Dundee. 47560 seemed to be a regular in Scotland along with its Cardiff stablemate 47558 May*flower*, which I had to Arbroath on 1A69 18.20 Carstairs–Aberdeen.

Sunday 9 February 1986. It turned out to be a long wait for 1E43 20.35 Aberdeen–King's Cross at Arbroath, as 47550 *University of Dundee* rolled in 120 minutes late! It was replaced at Edinburgh by 47475 as the train reversed to run via Carlisle. Another reversal at Newcastle saw 47402 *Gateshead* leave 78 minutes late, but further diversions via Leamside, Lincoln and Hertford North saw arrival in London 100 minutes late. Fortunes improved with 47620 *Windsor Castle* on 1F14 12.15 Paddington–Oxford, which although booked for a Class 47 could just have easily been a '50'. From Reading, erstwhile Inverness resident 47472, which had not gained a grey roof during an 18-month stay at Stratford and now a Gateshead machine, was taken to Birmingham on 1M03 11.59 Portsmouth Harbour–Manchester, and is seen waiting to be removed from the train for a Class 86 forward.

Wednesday 12 February 1986. An interview at Newcastle University prompted an early start on the 06.50 unit from Derby to Nottingham for 47561 to Sheffield on 1S49, an HST to York for 47444 *University of Nottingham* on 1E94 08.05 Liverpool–Newcastle. There were clearly problems at Newcastle as 47444 then ran round the stock to form 1M73 11.20 back to Newcastle with a late departure. After the interview, I headed north with 47475 on 1N17 17.18 Newcastle–Berwick, before an overnight in Scotland. 47444 would take me south the next day on 1V85 09.23 Newcastle–Penzance.

Saturday 15 February 1986. Another Saturday, another trip to Scotland. The 1S79 22.15 King's Cross–Aberdeen had been 47561 to Newcastle, for 47604 forward. 1S79 did not officially call at Edinburgh, but it did stop there for a crew change, and no one seemed to mind if you got off. The trouble was that it was about five o'clock in the morning, not great on a freezing February morning! Some people did the first 'shove' to Haymarket for the internal overnight back in, but it just made the Moves Book look messy and departure from Edinburgh was behind 47053 on 1H05 07.06 to Inverness. At the Highland Capital, recently converted from 47083 (page 7) but now allocated to Inverness, 47633 awaits departure with 1T30 12.30 Inverness–Glasgow.

There was a far more Scottish feel to the traction back from Perth with 47461 *Charles Rennie Mackintosh* at the helm on 1H13 13.35 Glasgow–Inverness. 47461 had been erroneously outshopped from Crewe in InterCity colours complete with red stripe in August 1985, but it had been repainted into ScotRail livery at Inverness during January while the loco was on a C exam. It was the only Class 47/4 to sport ScotRail livery with the Caledonian blue stripe. Back at Inverness, 47049 was had 'over the top' on 1A58 17.32 Inverness–Aberdeen for a marathon night out on 1E43 20.20 Aberdeen–King's Cross. The train left the 'Granite City' behind 47524, which was piloted by 47209 from Dundee to Edinburgh. Here, the train reversed and departed for Newcastle via Carlisle behind 47544 at 23.50. Another loco change and reversal saw 47404 *Hadrian* depart south on the main line at 03.35, but the loco was on fire as it arrived into York! Its replacement was 47472, which worked to London via Knottingley, Lincoln and Hertford. Arrival into London was about an hour earlier than the previous week, so the safest option was the 11.30 HST to Reading, rather than hang about in London only to find a Class 50 on 1F14. The reward was Eastfield's 47562 *Sir William Burrell* on 1M00 10.52 Gatwick–Liverpool to Birmingham.

Sunday 16 February 1986. Stratford's 47457 *Ben Line* rolled into Birmingham on 1V84 11.15 Liverpool–Penzance, pictured at Bristol, where it was booked to be detached. 47457 had been ex works from Crewe on 9 January and was named at Southampton Central just three days prior to this photo. Surprisingly, given its condition, it would be transferred to Gateshead in May. Another ex-works loco took me home, 47638 on 1E37 12.45 Penzance–Newcastle, which was just four days out of works following its conversion from 47069.

Tuesday 18 February 1986. Another university interview, this time at Heriot-Watt in Edinburgh and the local education authority had generously offered to pay for a sleeper, so it seemed like a good opportunity to get 1S77 23.35 King's Cross–Edinburgh, which was sleepers only. 47500 *Great Western* was my ride to London on 1V16 17.25 Manchester–Paddington, and 47522 was 1S77 throughout. Arriving at Edinburgh at 07.05, which was far too early for anyone at a university to be up and about, there was time for a fill in move to Leuchars and back with 47018 on 2J25 08.15 Edinburgh–Dundee, for 47017 back on 2J08 09.30 ex-Dundee. At about 08.00, 47538 is seen with the stock for the 08.55 Edinburgh–Aberdeen.

Saturday 22 February 1986. At just before midnight on the Friday night, I had been standing at York watching 47416 approaching on 1S70 20.25 King's Cross–Aberdeen when it came to a halt under Holgate Bridge. This was the last train that 47416 worked as it was withdrawn to keep the other Generators running. Its indignity was intensified as it was hauled into York by a Class 08 before 47419, which had worked in on 1E31 14.13 from Portsmouth Harbour, replaced it. I decided to hang back for 1S79 22.15 King's Cross–Aberdeen, which was 47561 to Newcastle for 47413 forward. Bailing in the early hours at Edinburgh, 47210 was atop 1H05 07.06 to Inverness, seen here being shunt released, and 47426 was on 1T30 12.30 Inverness–Glasgow, seen here at the picturesque station of Dunkeld.

Heading back north from Dunkeld was 47541 *The Queen Mother* on 1H13 13.35 Glasgow–Inverness where there was the welcome sight of 47003 on 1A58 17.32 Inverness–Aberdeen. 47541 was enjoying a brief spell in Large Logo livery, at the time, following its repaint at Inverness during a D exam in August 1985. It was stopped on 7 March for Intermediate overhaul at Crewe and repainted in ScotRail livery in June 1986. Another long stint on 1E43 saw 47522 to Newcastle via Carlisle for 47401 *North Eastern* to King's Cross via Knottingley, Lincoln and Hertford.

Wednesday 26 February 1986. Engineering work at Proof House Junction meant that cross-country trains from the northeast were not calling at Birmingham New Street. 47477 on 1O86 14.00 Leeds–Brighton makes its booked stop at Burton, after which its next calling point will be Solihull, as it works the train throughout. Coming back north, 47513 *Severn* on 1E31 14.13 Portsmouth Harbour–York also worked throughout, but rather than calling at Solihull, it stopped at Small Heath instead, necessitating a DMU to be taken between the Birmingham suburbs. Note the 'Skol' and 'Ind Coope' brewery signs and the seeming absence of a driver!

Saturday 1 March 1986. The shape of things to come was evident with 47619 on 1H05 07.06 Edinburgh–Inverness, in place of the expected steaming '47/0'. Even 'the circuit' was being electrically heated with 47411 on 2J23 07.15 Edinburgh–Dundee, and 47558 May*flower*, which might as well have been allocated to a Scottish depot given the amount of time it spent north of the border, on 2J34 06.30 Dundee–Edinburgh. 47421 *The Brontës of Haworth* is seen running into Edinburgh on 5A55 from Craigentinny, working 1A55 08.55 Edinburgh–Aberdeen, which, in the absence of any steam heat, was taken to Aberdeen. Its running number belied the fact that, as D1520, it had been the first built of the 'standard' Class 47s, delivered from Brush to Finsbury Park on 5 June 1963.

The granite buildings in Aberdeen were generally grey, as was the weather. Trying to blend in with its fading grey roof, Stratford's 47570 has backed onto the stock of 1A55 to form 1B32 12.00 back to 'Auld Reekie'. The driver keeps a watchful eye as the shunter ties the loco on for the 131-mile run. Back at Edinburgh, there was finally some steam heat on offer with 47152 on 2J07 15.20 Edinburgh–Dundee, but ETH was dominating 'the circuit'; 47711 *Greyfriars Bobby* had been the 13.15 ex-Edinburgh and 47533 the 14.15. Earlier in the day, 47508 *Great Britain* had worked 2J21 05.55 Edinburgh–Dundee, which was booked for air-conditioned stock as it was the Dundee portion for the Glasgow–Poole.

Sunday 2 March 1986. Deputising for a Class 50, 47513 *Severn* rumbles into Oxford with the ECS for 1F11 16.40 to Paddington. The overnight had seen 47409 fail at Carlisle and be replaced by 47586. The booked engine change at Newcastle had then seen original InterCity-liveried 47487 work to King's Cross via Leamside, Knottingley and Hertford North. 47513 had earlier worked 1F01 08.50 Oxford–Paddington, and 1F14 back to Oxford at 12.15. 47412 would take me home on 1E20 13.22 Poole–Newcastle, which ran round at New Street to get it back to Gateshead for exam.

Saturday 8 March 1986. 47195 had been released from overhaul at Crewe on 6 January, sporting the correct blue livery for a boiler-fitted loco. It was, however, soon back in works for rectification and it was Valentine's Day before it was released again. It was the last Class 47 to retain an operative boiler after classified repair, but it was of limited use in England, although the Scottish Region were happy to put it to good use. The still gleaming loco is seen at Stirling on 1H05 07.06 Edinburgh–Inverness, but, sadly, it was removed at Perth, as the boiler was required to heat a Glasgow to Edinburgh football special and 47632 worked north in another glimpse into the future. It looks like the secondman is having a drive while the driver reads the paper.

Hybrid liveried 47604 blasts away from Dunkeld with 1T30 12.30 Inverness–Glasgow. 47604 reappeared on 1H13 13.35 Glasgow–Inverness, which had clearly started at Perth, giving me quite a long stand at Dunkeld. 47604 was taken back to Aviemore to drop into required 47635, which had finally made its way to its new home in Scotland, on 1B36 16.30 Inverness–Edinburgh. At Aviemore, there was just time to dash to the chippy on the station for an overpriced haggis supper. I should not have bothered; the haggis was awful, tasting something like you would imagine a boiled sock would. I got some stick this day for not doing the Edinburgh–York 'adex' with 47191, which most people had done for some 400 steam heat miles.

Sunday 9 March 1986. The 1E43 20.20 Aberdeen–King's Cross had been 47409 to Newcastle for 47419 forward via the same route as the previous week, but arrival in London was much more punctual, allowing a move to East Croydon rather than an HST out of Paddington to intercept 1M00 10.52 Gatwick–Liverpool. 47086 had been stopped on the night of 12 November and arrived at Crewe Works for overhaul and ETH fitment at 11.30 on 3 December 1985. It was released on 6 March as an Eastfield loco and worked 1O86 14.00 Leeds–Brighton forward from Birmingham the next day. It powered 1M50 08.55 Brighton–Manchester–Birmingham the next day, before returning south on 1O86 again. It retained its *Colossus* nameplates for a couple of months longer before the Scottish Region removed them and the loco later carried the less charismatic *Fife Region* plates. It carried another two names in its later career.

Saturday 15 March 1986. Following its trip to York on the previous Saturday, Crewe's 47191 had been purloined by the Scottish Region by virtue of its boiler. It had spent the Monday on 'the circuit', then worked 1H05 07.06 Edinburgh–Inverness on the Tuesday and was allocated 1B34 14.30 return but failed at Inverness with low oil pressure. Spending Wednesday and Thursday at Inverness for camshaft and timing gear repairs, Friday saw it working 'over the top' on 1A46 08.32 to Aberdeen, and it is seen here at Aberdeen, having worked the same train again, before running round to form 1H29 11.40 back. Also working 'over the top' on this day were 47527on the 07.40 ex Aberdeen, 47053 on the 09.40 and 47550 *University of Dundee* on the 13.40.

Stabled at Aberdeen were 47407 *Aycliffe* off 1S79 and 47426 for 1B32 12.00 to Edinburgh, along with 47546 *Aviemore Centre*, which had worked 1A67 17.25 Glasgow–Aberdeen the previous night, and Haymarket's 47003. The latter had enjoyed a decade at Stratford from May 1975 to May 1985 but had been made redundant when steam heat finished in East Anglia. It had originally moved to Gateshead, but three months later continued north to Scotland, where its boiler could once again be put to good use. 47546 would undertake a rare trip on the Far North Line five days later when it worked 6H25 Millerhill–Muir of Ord.

Crewe's 47205 waits to leave Dundee on 2J04 13.30 to Edinburgh. Other steam heat locos on 'the circuit' on this day were 47209 on the 12.15 ex-Edinburgh, 47004 on the 13.15 ex-Edinburgh, while 47152 had worked the 10.15 earlier, but was replaced by 47631 for the 15.20. The 14.15 ex-Edinburgh was 47619. The 07.06 Edinburgh–Inverness had also been ETH with Stratford's 47596 *Aldeburgh Festival*.

Sunday 16 March 1986. 1E43 on Saturday night was 47403 *The Geordie* from Aberdeen to Newcastle, where 47596 *Aldeburgh Festival* was waiting to take over for the run to London in the early hours of Sunday after its trip to Inverness on the Saturday. Arrival in London was not prompt enough for another trip to the delights of East Croydon, so the 11.30 HST from Paddington to Reading had to suffice. Inverness's 47586 had once again been to the South Coast and it was the power for 1M00 10.52 Gatwick–Liverpool, seen here at a rainy Birmingham International.

Going in the opposite direction was 47427 on 1O74 11.40 Manchester–Gatwick, seen at Oxford. 47427 had returned to Gateshead on 2 February after being part of a batch of Eastern Region Class 47s that were loaned to the London Midland, with various others going in the opposite direction, as the Eastern had been far quicker in fitting speed sensors for One Man Operation than the Midland had been. However, on this day it was transferred again and was now a Tinsley loco! It arrived at Crewe for what would be its final overhaul at 09.40 on 28 May, with the work being completed on 22 August.

Friday 21 March 1986. Derby to Inverness, the long way round. 47642 worked 1O19 08.00 Newcastle–Poole from New Street, and is seen after running round at Reading, although not booked to do so. Officially renumbered from 47040 just ten days previously, it had worked 1V74 08.28 Leeds–Cardiff on the 12th. It spent the next day undergoing special tests at Canton and was at Llandarcy on an oil train on the 14th. It worked 1O04 06.43 Exeter–Waterloo on the 19th, before taking 1E31 14.13 Portsmouth Harbour–York to Birmingham as booked. It re-engined 1O03 06.50 Liverpool–Poole at Coventry on the 20th and had worked 1M50 08.50 Brighton–Manchester on this morning.

Saturday 22 March 1986. After a long night on 1S79 22.15 King's Cross–Aberdeen throughout (47527 to Newcastle, 47631 forward), it was off in chase of steam heat again. 47461 *Charles Rennie Mackintosh* was taken to Elgin on 1H27 09.40 Aberdeen–Inverness to intercept 47049 on 1A48 10.32 Inverness–Aberdeen back to the 'Granite City'. On its way back west on 1H31 13.40 Aberdeen–Inverness, 47049 pauses at Elgin waiting for 47461 to clear the section as it returns on 1A52 14.32 Inverness–Aberdeen.

On a day that was beyond dreich at the Highland Capital, 47430 waits departure on 1B36 16.30 Inverness–Edinburgh, while 47049 has been shunt released in order to work back east again. 47430 had been allocated to Haymarket in September 1985 and so, as a Scottish loco, it was released from overhaul at Crewe on 15 January 1986 in ScotRail livery. Another torturous East Coast overnight lay ahead. Having got off 47463 on 1C79 20.10 Aberdeen–Carstairs at Kirkcaldy, in order to go to the chippy to get a haggis supper to eat on 1E43, the haggis was a distant memory before 1E43 finally arrived 80 minutes late behind 47214 and an ailing 47480 *Robin Hood*. 47640 went forward from Edinburgh to Newcastle via Carlisle before 47448 took over for the run to London, where it arrived 40 minutes late at 10.45.

Sunday 23 March 1986. Over at Liverpool Street, 47579 *James Nightall GC* departs with an additional service to Harwich Parkeston Quay, while 47605 waits time with 1H10 12.35 to Cambridge. I made my way to Gatwick for 47643, which seemed in no hurry to get to Scotland, on 1M15 14.45 to Manchester.

Thursday 27 March 1986. Although booked for a Gateshead Class 47, 1E54 06.50 Paddington–Hull had a nasty habit of producing a Class 45, as the loco started the day by working 1O62 06.02 Derby–Brighton to Birmingham. Or, even if 1O62 was a '47', it would sometimes run round at Birmingham and a spare Class 45 would be kicked out from Saltley for 1E54. On this day, 1E54 had got a '47', if not a very inspiring one, as 47629 had worked to Hull and run round its meagre four coach load, the rear five coaches of 1E54 having gone to Leeds. 47629 waits to work 1J01 13.22 Hull–Sheffield, the portion for 1O86 14.00 Leeds–Brighton. It will later work 1L18 18.00 Sheffield–Leeds, and was then booked on 1A41 vans to King's Cross.

The disappointment of 47629 turned to dismay with the appearance of 47632 on 1E36 09.24 Dundee–King's Cross relief, for the run to London. Fortunes improved after crossing to West London with 47560 *Tamar* on 1B56 18.07 Paddington–Hereford, which was enjoyed to Oxford for 47468 to Birmingham on 1M40 17.05 Poole–Liverpool.

Saturday 29 March 1986. Long-time Western Region loco 47246 was stopped for overhaul in the early hours of 9 December 1985 and arrived at Crewe Works at 11.25 ten days later. The loco was reallocated to Eastfield while under conversion to 47644, but because it had not entered works as a Scottish loco, it was painted into Large Logo livery, rather than ScotRail. It had only been released from works at 17.00 on 27 March and is seen here at Wellington with 1J25 11.40 Euston–Shrewsbury, which it had taken over from 86428 at Wolverhampton. There was only time to take 47644 to Wellington as engineering work meant that 1O92 13.52 Liverpool–Portsmouth Harbour was being diverted via the Golden Valley. So, it was a case of having to catch a DMU back to Wolverhampton for 47444 *University of Nottingham* to Reading on 1O92 for 47643 back to Birmingham on 1M40 17.05 Poole–Liverpool via the West Coast Main Line.

Wednesday 2 April 1986. Following right behind 47578 on 1O74 09.58 Manchester–Brighton (page 3) was 47508 *SS Great Britain* on the Calvert to Bristol 'Avon binliner'. A most unusual working for one of Old Oak Common's flagship locos.

Thursday 3 April 1986. It was a good day for Gateshead locos on 'The Line', as the locals called the route from Birmingham to Reading. 47403 *The Geordie* had worked 1O74 09.58 Manchester–Brighton forward from Coventry, 47407 *Aycliffe* was on 1M23 14.40 Poole–Liverpool to Birmingham, and later in the day, 47525 was atop 1M40 17.05 Poole–Liverpool to Birmingham. 47528, the last Class 47 to be built at Crewe, entering service on 18 February 1967, waits to get back underway at Banbury with 1O92 13.52 Liverpool–Portsmouth Harbour, which it will work throughout, after starting the day on 1M10 06.03 Paddington–Liverpool.

Saturday 12 April 1986. The erstwhile 1M12 16.40 Nottingham–Barrow had been superseded by the 14.10 Cambridge–Blackpool North, which kept the same reporting number. It was booked for a Crewe '47', which worked into East Anglia on 1E86 07.20 Blackpool–Cambridge the previous day and was then used on fill-in turns on the Great Eastern. 47552 had performed an extra duty on this day when it had been sent to Elsenham to rescue a failed 47420 on 1H16 11.35 Liverpool Street–Cambridge prior to working 1M12. 47552 would arrive at Crewe Works for Intermediate overhaul and twin tank fitment on 6 May.

Saturday 19 April 1986. The pioneer of the class, 47401 *North Eastern* makes its booked stop at Water Orton with 1G14 07.17 Derby–Birmingham. 1G14 was a curious train, booked to call at all stations, including Wilnecote at 07.47 and Water Orton at 08.00, giving plenty of time for a photo before finally arriving at New Street at 08.15, after a very leisurely 41¼ mile journey. The loco was then booked to work 1O27 06.45 York–Portsmouth forward as part of a Bescot diagram so, unlike 1O62 off Derby, it was almost invariably a Class 47. On this day, however, 47401 did not work 1O27, but later worked 1O86 14.00 Leeds–Brighton to the south coast.

47418 was on GD16 diagram, which started on 1O62 06.02 Derby–Brighton to Birmingham, followed by 1E54 06.50 Paddington–Hull to Humberside, seen here at Chesterfield. The train was booked for a Class 50 to Birmingham, but on this occasion had been worked in by 47606 *Odin*. The 1O19 08.00 Newcastle–Poole was booked for a Class 45 at the time, but 47472 went south on this, as I waited for required 47645 on 1S85 07.17 Harwich–Glasgow. 47645 had been released from Crewe on 19 March, following conversion from 47075.

47509 *Albion* was on GD28 diagram, 1V85 09.23 Newcastle–Penzance, which it would work to Plymouth. Thankfully, 47472 returned north on 1E63 09.40 Poole–Newcastle. Kingmoor had received an allocation of the class in September 1985 in order to take the load off Crewe and one of these, 47478, runs into Platform 2 at Sheffield with 1E87 11.15 Glasgow–Harwich, which will run round here and will be a long way from its Carlisle home by the end of the day.

Saturday 3 May 1986. The 1O62 06.02 Derby–Brighton was worked to New Street by 47524, where another Gateshead loco was waiting to take over for the run to the south coast. The driver of 47473 looks back impatiently to see what the delay is in the unloading of the mail.

The Gateshead theme continued with the reappearance of 47526 on 1M88 06.33 Poole–Manchester, which it will work to its destination. Seen here blasting out of Coventry in a reversal of the previous night, when it had worked 1O23 to Reading for 47446 forward, it had taken over from 47446 at Reading. 47526 had worked 1E53 13.00 Bangor–York, on 28 January, and later worked 1M41 22.34 York–Shrewsbury TPO, before running light to Crewe. By 11.15 it had arrived at Crewe Works for what would be its final overhaul. It was released in the second week of April and worked 1V04 07.05 Holyhead–Cardiff forward from Crewe on the 11th and was allocated 1M86 14.12 Cardiff–Holyhead back, but seems to have had trouble as this was worked by 47432, while 47526 later worked 1M82 20.00 Cardiff–Crewe. It was in trouble again on 24 April when it failed at Southampton on 1M88 06.25 Poole–Manchester, and 47304 was used to take the train to Reading. Yet another Gateshead loco was next up, as 47429 rolled in a few minutes later on 1O27 06.45 York–Portsmouth Harbour, which had been brought into Birmingham by stablemate 47525.

Monday 5 May 1986. The magnificent sight of a Class 47 on a rake of MGRs, as 47233 thunders through Oxford after discharging its load at Didcot CEGB. The '47s' had previously dominated the Didcot MGRs but had in the main been replaced by Class 56s and Class 58s by this time. 'Bicester' has obviously got a similar photo!

Sunday 11 May 1986. It was rather unusual to have a choice of Class 47s at Derby. In Platform 6, 47466, out of works for less than a week, stands in for a Class 45 on 1C26 15.45 Derby–St Pancras, while 47636 is in Platform 4 with 1V87 10.30 Newcastle–Penzance. With hindsight, a run to London with 47466 would have been the better move, as it later worked back on 1P24 19.45 St Pancras–Derby, but it would have been brave as 1P24 could easily have been a Class 45.

Eastfield's 47636 *Sir John De Graeme*, which had only been named on 28 April in a ceremony at Falkirk Grahamston station, waits time with 1V87 10.30 Newcastle–Penzance, which it will work to Plymouth. 47636 did not work back on 1E91 09.33 Penzance–Newcastle the next day, remaining instead on the Western Region. It was involved in a SPAD (signal passed at danger) on the 13th after it took over 1V38 08.10 Portsmouth Harbour–Cardiff at Bristol and passed a red signal at Filton, requiring it to go for brake and speedo tests at Cardiff Canton.

The day was the end of an era, with many famous trains running for the last time. Two of these, 1O92 15.21 Leeds–Portsmouth Harbour and 1E08 16.35 Birmingham–York, had regularly been steam-heated up until the demise of this form of heating, although neither was a regular Class 47 train. 1O92 was 47477 throughout today, while 1E08 was in the hands of 47564. The experiment of cross-country trains serving Hull also finished today with 47515 working 1V86 10.54 Hull–Cardiff from Sheffield, while 47525 worked the final 1E79 18.20 Cardiff–Hull.

It was a big day for reallocations as locos were moved around for the summer timetable. Amongst others, Stratford lost 47457/458/566/568/569/587 to Gateshead, which in turn lost 47425/426 to Tinsley and 47450/472/473 to Bescot. The LMR eased the burden on Crewe Diesel with the movement of 47339/431/432/588/610 to Bescot. North of the border, there was a shuffling of the pack with many Class 47s transferring between Eastfield, Haymarket and Inverness, as Haymarket lost its '47/4s' and gained more boilered locos. One transfer that would pay almost immediate dividends was that of 47085 which, along with 47146, moved from Stratford to Cardiff.

Chapter 3

Summer 1986

Below: Saturday 17 May 1986. Despite the atrocious weather, note all the umbrellas, summer got off to an excellent start with 47085 *Mammoth* turned out for 1E72 10.30 Birmingham–Yarmouth to Norwich, seen here in the rain at Leicester. 47085 had returned to its Western Region roots with a move from Stratford to Cardiff Canton the previous Sunday, but would be reallocated again the next day to Bristol Bath Road. Taken to Thetford, this was my 479th class member.

My 479th Class 47 was followed swiftly by the 480th with Thornaby's 47346 on 1M68 12.45 Yarmouth–Birmingham. By having a lie-in and starting the day with 47627 *City of Oxford* on 1O05 06.52 Leeds–Poole to Birmingham, I had missed 47346 on 1E66 08.19 Birmingham–Yarmouth, but not to worry as it was booked to return and is seen here at a rainy March.

Sunday 18 May 1986. Having deposited the wagons of a ballast train in Chaddesden Sidings, Tinsley's 47374 runs past Derby Locomotive Works on its way for servicing at Derby No 4 Shed. 47374 had been overhauled the previous November but had not worked the famous test train to Holyhead, leading to much disappointment. It would transfer to Stratford in June, which was not going to make it any more regular on passenger trains. Perhaps not surprisingly, it would go on to be one of my last Class 47s for haulage.

Saturday 24 May 1986. This was a bit of a surprise. After 47528 to Birmingham on 1O05, 47560 *Tamar* was on 1E72 in place of the expected Class 47/0 or 47/3. Now, I like a bit of *Tamar* as much as the next man, but it was not what you were after on a summer Saturday. Fortunately, Stratford's 47096 was heading west on 1M67 10.50 Yarmouth–Birmingham, in place of the booked pair of Class 31s, so a gang of us found ourselves at Peterborough for 47096.

47096 returned from whence it came on 1E80 16.21 Birmingham–Norwich, seen here at Nuneaton, prior to returning to Birmingham with former Class 48 47115 on 1M68, and, with that, No 481 was in the book. 47115 was then had back to Leicester on 1E82 18.25 Birmingham–Norwich for 47560 back to New Street on 1M76 16.00 Yarmouth–Birmingham, on what was becoming a standard Saturday move.

Sunday 25 May 1986. Spring 1986 had seen Tinsley get an allocation of Class 47/4s for the first time when 47427/428/438/439 arrived in March. From the May timetable, the Harwich–Glasgow and Sunday Harwich–Manchester were booked for a Tinsley '47' between East Anglia and Sheffield. Somewhat ironically, 1M74 07.20 Harwich–Manchester Piccadilly and 1E87 15.20 return had produced Stratford loco 47582 *County of Norfolk* today, seen here at Nottingham, to and from Sheffield on this date.

Bank Holiday Monday 26 May 1986. Class 47s were still rare on railtours at this point, but Hertfordshire Railtours' 1Z27 06.25 King's Cross–Matlock 'Peak Explorer' featured 47445 throughout. Despite having previously been the main line from Derby to Manchester, what was now the Matlock branch was very rare territory for loco-hauled trains. With the single line token from Ambergate used to unlock the groundframe at Matlock, the train was able to enter the yard to run round and the groundframe was locked behind it to allow the regular DMU service to access the branch. Having run-round, 47445 waits in the yard to draw into the station to form 1Z27 17.40 Matlock–King's Cross.

Saturday 31 May 1986. The Yarmouths continued to produce Eastern Region rarities and Thornaby's 47361 *Wilton Endeavour* was about as rare as they came. 47361 had been turned out on 1E72 with stablemate 47346 on 1E66/1M68/1M82 again. After an earlier trip to Thetford, when it became No 482 in the book, 47361 waits time at Leicester on 1M76.

Thursday 5 June 1986. Another of the elusive Thornaby Class 47s, 47363 *Billingham Enterprise* rests on Toton in the company of local residents, 20188 and 20101. Like its stablemates, 47363 was exceptionally rare on passenger trains and today had probably worked 6M38 06.25 Port Clarence–Long Eaton tanks, and was waiting to work back on 6E50 at 16.55, this being staple work for a Thornaby '47'.

Saturday 7 June 1986. The early morning gen call came through that Thornaby's 47305 was working 1E07 07.10 Sheffield–Skegness, and, seeing as I only had a rather embarrassing 1.24 miles off it previously, a trip to 'Skeg Vegas' rather than Thetford seemed in order. After quite long enough at the 'resort', 47305 worked back on 1E02 10.43 Skegness–Sheffield, seen here at Boston, with the cab detailing of the creative Thornaby staff seen to good effect.

Getting back to Derby, 47567 *Red Star* was on 1V65 09.50 Newcastle–Penzance, with 47407 *Aycliffe* on 1O11 09.50 Glasgow–Poole out of Birmingham. The sensible move would have been to take 47407 to Banbury for 47230 back on 1M23 13.57 Poole–Manchester, but for some reason I got off eight miles later and had 58037 back to Birmingham on 1M14. Going south again with 47620 *Windsor Castle* on 1V81 12.55 Llandudno–Paddington, 47230 was made at Leamington for my one and only run off it (No 483). 47230 was withdrawn the following January, having spent its entire life at Cardiff Canton.

The Yarmouths on this day were 47195 on 1E66/1M68/1E82, seen here at Leicester on 1E82, and 47131 on 1E72/1M76, which was had back to Birmingham. 47131 would be another early casualty, it was derailed and turned onto its side at Dorrington while working 6M27 Waterston–Albion on 19 February 1987 and was considered beyond economic repair. 47195, on the other hand, had been overhauled in January and would go on to put in sterling work for the Petroleum sector in years to come.

Friday 13 June 1986. The withdrawal of 47405/414/416 earlier in the year had highlighted the fact that any of the Generators that had not been through works recently were vulnerable, and, having last been overhauled in November 1982, 47419 fell into this category. Seen on arrival with 1M03 08.05 Portsmouth Harbour–Liverpool, 47419 lasted in traffic until January 1987.

47647 *Thor* had worked 1D43 10.00 Euston–Holyhead on test from Crewe Works, with 47527 going along as insurance, on 16 April. After minor rectification, the former 47091 had returned to traffic on 22 April, now based at Crewe. Here, it awaits departure on 1E74 14.03 Liverpool–Scarborough, which it will work to York. The stock had been brought in by 47526, which would then work 1G61 14.30 to Birmingham.

47128 had started the day working 1E00 07.03 Liverpool–Scarborough to York, then 1E06 09.03 Liverpool–Scarborough to the seaside. It then crossed the Pennines again with 1M26 12.53 Scarborough–Bangor, and the shunter has already jumped in between the loco and first coach for the loco to run round to work 1J31 19.20 to Manchester Victoria. Preferring not to hang around in this hostile North Wales outpost (everyone started speaking Welsh when you went into a pub here), I returned to England with 47623 *Vulcan* on 1G56 18.00 Holyhead–Birmingham. 47128 arrived at Crewe Works at 09.35 on 18 June for conversion to 47656.

Saturday 14 June 1986. A day that started at Holyhead with 47491 *Horwich Enterprise* on 1A04 01.15 to Euston. Liverpool was reached off 47523 to Oxford on 1O01 06.26 Wolverhampton–Poole, 47590 *Thomas Telford* back north on 1M88 06.33 Poole–Manchester, and 47555 The *Commonwealth Spirit* on 1M37 08.20 Paignton–Manchester to Stockport. Two former Stratford charges meet at Liverpool Lime Street; 47568, now domiciled at Gateshead, waits departure with 1E88 16.03 Liverpool–Newcastle, while 47122, now of Crewe Diesel, has been released after working in on 1M31 08.15 Yarmouth–Liverpool.

The last day for 47568 based at Stratford had been Sunday 11 May, when it had worked 1C74 13.13 King's Lynn–Liverpool Street. The next night, its first as a Gateshead loco, saw it replace the failed 47435 at York on 1S72 22.30 King's Cross–Edinburgh. It spent the first week of its new career hard at work on ECML overnight services, culminating with 1S70 20.12 King's Cross–Aberdeen to Edinburgh on the 18th. The next day saw it at the Highland Capital, working 1H07 09.23 Edinburgh–Inverness and 1B36 17.30 return, quite a change from its incarceration in the flatlands of East Anglia.

Back at Crewe and despite its lack of train heat, 47148 had stuck to diagram off 1K34 17.40 Llandudno–Crewe and waits in Platform 12 to work 1V07 20.48 Crewe–Cardiff, which was a Class 33 diagram Monday to Friday, but a '47' on Saturdays. Sadly, this was a move to nowhere and I had to let it go. Instead, I headed for another night in Holyhead with 47561 working 1D71 19.00 Euston–Holyhead forward from Crewe at 21.17.

Monday 16 June 1986. A nice sociable 10.19 start could be had by turning out for 1M55 08.35 Leeds–Birmingham, providing it was not a 'Peak'! This gave two options for a move to Liverpool, as the loco and stock off 1M55 went forward as 1F09 12.29 Birmingham–Liverpool, following 1M03 08.05 Portsmouth Harbour–Liverpool. 47440 was 1M55 today, with 47436 on 1M03. At Liverpool, 47602 *Glorious Devon*, which would be the engine for 1O46 15.10 to Portsmouth Harbour, sits in one of the sidings. Behind it stands 47532, which I presume had worked in on 1M67 09.53 from Scarborough and would therefore be the engine for 1G61.

Sat in another siding was required 47286 and I thought that I was going to have to pitch my tent until it worked. But no need, it was fired up and proceeded to drop onto 1G61 14.30 to Birmingham, which should have been the loco off 1M67. 1G61 then formed 1E78 16.52 Birmingham–Leeds, happily taking me back home with No 484 in the book. Bob Geldof may not have liked Mondays, but 47286 seemed to, as it had worked 1M26 12.53 Scarborough–Bangor the previous Monday.

Tuesday 17 June 1986. Taking 1M03 to Liverpool for 1G61 was becoming a regular move with 47602 *Glorious Devon* for 47528 today. However, at Wolverhampton there was no sign of an electric loco to work 1A63 15.33 Shrewsbury–Euston forward. Getting off on spec, the gamble paid off as Tinsley's 47438 arrived with 1A63 and continued forward. The worst fears of a loco change at New Street proved unfounded and remarkably the diesel worked through to London, where it is seen on the blocks after a few of us had enjoyed a journey in the 'egg-cup' seats in the buffet car. The only fly in the ointment was that 47438 was then allocated to work a Freightliner north, rather than working back out of Euston.

Unable to find a Class 87 out of Euston, I had to settle for 86247 on 1G44 19.10 Wolverhampton to Coventry for 47639 *Industry Year 1986*, which by coincidence had been named at Euston just eight days previously, to Birmingham on 1M40 17.05 Poole–Liverpool. 47639 was planned to be named in April but somewhat embarrassingly, given the name, the ceremony was cancelled due to industrial action! At Coventry, 47406 *Rail Riders* was sat in the naughty siding after failing earlier in the day on a Liverpool to Ascot 'racex' and being replaced by 47615 *Caerphilly Castle*.

Thursday 19 June 1986. 47610 was 1F09 today, followed by 47522 on 1G61 to Wolverhampton for 47558 May*flower* back to Liverpool on 1M14 13.15 Paddington–Liverpool. At Lime Street, 47420 was on the blocks after arriving with 1M75 13.53 from Scarborough. Any further plans for the day went out of the window with the gen that required 47223 was working 1K34 17.40 Llandudno–Crewe vice a Class 33, necessitating a desperate leap.

Cue a dash downstairs for a Mersey Rail electric to Hooton, then a DMU to Chester, which gave me what turned out to be a fairly comfortable 15-minute connection at Chester for 47223, one of the hard-to-get Immingham ones, to Crewe. The desperate leap was because of the worry that 47223 would be swapped out at Crewe and would not work the next leg of the diagram, 1V07 20.48 Crewe–Cardiff, but of course it did. Not fancying an overnight, I took 47598 back to Birmingham on 1V99 19.05 Liverpool–Paddington.

Saturday 21 June 1986. The day of the first Ian Allen Network Day with unlimited travel over Network Southeast for a fiver and the chance to go off the beaten track. 47547 had taken over 1O28 07.49 Liverpool–Dover Western Docks at Mitre Bridge and has run round at the port ready to depart with 1M04 13.45 back to Liverpool, which it will work to Willesden West London Junction. InterCity services from the northwest to Dover had been inaugurated in May 1986, but they never really took off as BR hoped and they were down to weekend-only workings from May 1988.

After a spin to Bishops Stortford with 47579 *James Nightall GC* on 1H26 16.35 Liverpool Street–King's Lynn for 47096 back on 1C87 17.05 Cambridge–Liverpool Street, 47577 *Benjamin Gimbert GC* waits to work 1H32 18.35 Liverpool Street–King's Lynn. 47577 and 47579 were named at March on 28 September 1981 to mark the bravery of the two men, whose story is told on the brass plaque below the nameplate.

Sunday 22 June 1986. Fresh out of the box! 47535 *University of Leicester* had only been released from Crewe Works on the evening of 17 June following Intermediate overhaul. It had spent the previous day on the Holyhead route, while here it has backed on to 1S63 12.45 Euston–Glasgow to drag it to Preston due to engineering work. 47535 was one of the few '47s' with flush fronts at both ends, following its collision with a DMU at Luton on Saturday 28 May 1983. Both cabs of the loco were crushed, and it was in works from 21 June until 27 January 1984.

47343 opens up at Crewe on 1V13 15.00 Manchester–Swansea. It had earlier worked 1M74 07.20 Harwich Parkeston Quay–Manchester from Sheffield, where it had replaced 47426, but rather than returning to Sheffield on 1E87 15.20 Manchester–Harwich, it had been put onto 1V13 instead. 47343 was reluctantly flagged in order to get home at a reasonable hour, with 47447 on 1O46 14.20 Liverpool–Portsmouth Harbour taken to Birmingham for 47603 *County of Somerset* home on 1E61 10.50 Penzance–Newcastle.

Saturday 28 June 1986. After a lacklustre start with 47532 on 1J18 07.25 Birmingham–Aberystwyth, the day showed a marked improvement with required 47323 on 1E84 07.05 Cardiff–Newcastle, with a rake of Mk.1 stock in place of the booked HST. So, No 486 was in the book with a run to York. There was no time for a photo at York as 47461 *Charles Rennie Mackintosh* was running into York on 1V65 09.50 Newcastle–Penzance at the same time as 1E84. Back at Birmingham, 47530 was had to Burton on 1E65 09.18 Newquay–Newcastle, followed by 47445 on 1E78 10.10 Penzance–Leeds to Derby. 47145 worked back with 47323's stock as 1V78 13.20 Newcastle–Plymouth, and is seen after running round at Gloucester.

Sunday 29 June 1986. The 1D54 09.30 Birmingham–Holyhead had become an interesting Sunday train; ostensibly booked for a Bescot Class 47/4, it had not got one since 1 June when 47645 *Robert F Fairlie* had done the honours. After that, 47234 had worked the train on 8 June, 47063 one week later and today 47099 was in charge. The loco off 1D54 was booked back on 1G12 17.58 Holyhead–Birmingham, but no one in their right mind fancied a Sunday in Holyhead, so the move was to get off at Bangor for 1A27 13.20 Holyhead–Euston back to civilisation. This conveniently dropped into 1O46 at Crewe and 1E61 at Birmingham as per the previous Sunday, which were 47513 *Severn,* 47448 and 47503 respectively.

Saturday 5 July 1986. Friday had been traumatic as 47245 had worked 1V76 13.18 York–Plymouth, just feet away from where I was working in the Travel Centre at Derby, but my supervisor refused to give me a flexible lunch break and the winner passed me by. Getting No 487 in the book in the form of 47652, which had only been officially renumbered from 47055 two days previously, on 1O07 09.02 Manchester–Poole from Stafford to Oxford was scant consolation. Later in the day, a short run from Birmingham to Derby on 47079 *GJ Churchward* was had on 1E65 09.18 Newquay–Newcastle.

Saturday 12 July 1986. Another rainy Saturday in the West Midlands sees 47220 at the head of 1J18 07.25 Birmingham–Aberystwyth at Wolverhampton, one of only three diagrams that were booked for no-heat '47s' in the Birmingham area, along with 1E66 and 1E72. Going to Shrewsbury with 1J18 was a move to nowhere, but it could be done to Wolverhampton with time to spare to get back to New Street for 1E66 if desired. 47350 was 1E66 on this day, but I was about to be sent into a tailspin by the gen that I had two winners out but could only get to one or the other of them.

The choice was between 47284, which was working on the Western Region, or 47310, which was on Blackpool shuttles. 47310 was based at Stratford at the time and was one of the rarest '47s' on passenger trains, whereas Bristol-based 47284 would probably be the more likely one to work in the future. An easy decision but, rather than keeping a cool head, I went into a flap to get to Preston and boarded 85026 on 1S53 07.26 Coventry–Glasgow to get to Crewe. Realising my error, there was time for a fill-in to Chester with 47437 on 1D07 09.14 Crewe–Holyhead out for 47561 back on 1A41 09.39 Llandudno–Euston. Had I thought about it, I could have had 47450 back to New Street on 1O03 07.10 Liverpool–Poole for 1M10 06.58 Paddington–Llandudno to Crewe. The error was compounded as there was a '47/0' on the train in place of the booked '47/4'! 47051 is seen blasting out of Crewe on 1M10.

87020 *North Briton* was on 1P18 08.55 Euston–Blackpool North–Preston, the train that 47310 should work forward if it stuck to diagram. At Preston, 87020 was unhooked and disappeared northwards. A nervous wait ensued before a no-heat, grey roofed '47' appeared and was coupled onto the train. The brake test was undertaken, and we were on our way. No 488 was in the book! After being shunt released at Blackpool, 47310 stands at the head of 1A62 14.16 to Euston, which it will work to Preston for 86210 *City of Edinburgh* forward. Of course, time would prove that I had made the wrong choice today as 47310 went on to become quite common on passenger trains after this, while 47284 proved elusive.

The sands of time were running low for 47403 *The Geordie*, which had started the day working 1S49 07.15 Nottingham–Glasgow to Preston, before taking 1M21 08.50 Glasgow–Blackpool North to the seaside, where it was a grey day indeed. Here it stands at the head of a rake of Mk.1 stock forming 1S68 14.39 back to Glasgow, which it will work to Preston. 47403 had just two months left in traffic at this point, but sister loco 47408 *Finsbury Park* was even closer to the end, as it worked its last known passenger train on this day.

Sunday 13 July 1986. After another diabolical overnight involving very little sleep and the delights of 47563 and 47449, 21 miles on 47359 seemed like scant reward. Having been to Poole the previous day, 47359 had worked 1A06 08.56 Wolverhampton–Euston to Birmingham before taking 1O29 09.03 Wolverhampton–Dover Western Docks to Nuneaton, where the driver has already eased up for uncoupling. This should have been a 'drag', with the electric loco to work forward to Mitre Bridge being hauled dead along the non-electrified route from Birmingham, but clearly this has not been the case. Note the stub of the through steam pipe on the bufferbeam. The Class 47/3s were all built with through steam pipes, despite never having boilers.

47470 *University of Edinburgh* is heading far from its Eastfield home on 1O46 14.20 Liverpool–Portsmouth Harbour, seen here at Crewe awaiting its booked departure time of 16.02, having already had a tour of Cheshire on its way from Merseyside. 47470 was enjoying its summer holiday south of the border; having worked 1E86 07.20 Blackpool–Harwich to Sheffield on Saturday 5 July, it worked 1S49 07.15 Nottingham–Glasgow to Preston on Friday 11th, and had worked 1E86 to Sheffield again on the 12th.

Saturday 19 July 1986. As had happened the previous Saturday, a Class 47/0 had been turned out on 1M10 06.58 Paddington–Llandudno. 47017 was halfway through a four month stay at Eastfield in between stints based at Haymarket. It was reached at Banbury off 47466 on 1O01 06.26 Wolverhampton–Poole and, with 47162 and 47113 on the Yarmouths, a trip to Llandudno rather than Thetford was in order. Here, the Scottish machine shunts the empty stock into the sidings to run round. No air-conditioned comfort in first class today!

Run round and ready to go, 47017 awaits departure of 1V81 12.55 to London Paddington, which it will work to Birmingham. The train was booked for a Class 50 from Birmingham, but there was a bonus today as 47613 *North Star* took over at New Street. In the background, 47526, now sporting the *Northumbria* nameplates from the withdrawn 47405, is on 1E53 13.12 to York. 47613 was later taken to Leamington for 47438 back to Birmingham on 1M23 13.57 Poole–Manchester for 47162 on 1E82 to Leicester and 47113 on its return to New Street on 1M76.

Sunday 20 July 1986. It had now become the norm to do a torturous overnight involving Manchester, Stafford, and Crewe as 1D54 09.30 Birmingham–Holyhead was regularly getting non-ETH '47s' and there was no way to it from Derby. 47599 had been 1A01/1H02 and 47619 was 1G00 on another night of little sleep and the reward was 47201 on 1D54, seen here as it paused at Llandudno Junction. Getting back on to Bangor, 47417 was 1A27, 47433 1O46 and 47612 *Titan* 1E63 13.25 Poole–Newcastle forward from Birmingham.

Sunday 27 July 1986. I had to work an early shift on the Saturday, something that was made more palatable when I learned that 1E66 was a pair of Class 31s. Finishing in time to catch 1E22 10.25 Portsmouth Harbour–Leeds from Derby at 15.08 to Wakefield with 47489, an HST was taken to Peterborough to put me in position for what I thought was going to be 47357 on 1M76 16.00 Yarmouth–Birmingham. What I did not know was that 47357 had failed at Syston on 1E72 and had been replaced by a Class 31, which to my horror now came pedalling in on 1M76. Bowled! A night out in Leamington beckoned, so 31308 had to be suffered to Birmingham for 47484 *Isambard Kingdom Brunel* on 1V99 19.05 Liverpool–Paddington. The Sunday morning saw a drive to Rugby for 1M34 06.25 Newhaven Marine–Manchester, which was worked to Birmingham via Nuneaton by 47411.

Inverness allocated 47633 backed onto the rear of 1M34 at Birmingham to take the train to Manchester via Bescot, seen here departing Stoke in all its Highland finery. 47633 had worked 1O12 16.40 Liverpool–Poole to Reading two days previously and 1M10 06.58 Paddington–Llandudno on the Saturday, before taking 1V81 12.55 Llandudno–Paddington back to Birmingham. Tuesday 29 July saw it on 1E88 16.03 Liverpool–Newcastle, before working 1A40 20.54 to King's Cross. By the night of Thursday 31st, it was on its way home on 1H01 23.23 Edinburgh–Inverness.

Saturday 2 August 1986. With 1V32 23.50 Glasgow–Penzance running extremely late, 47189 and the 'raspberry ripple' rake of Mk.1 stock were chucked out from Oxley to work 1Z32 05.36 Wolverhampton–Penzance, which it would work throughout in its path. 47189 blasts out of Cheltenham, where a short wait would ensue for 47474 back to Birmingham on 1M49 05.30 Plymouth–Liverpool, which was conveniently loco-hauled vice HST.

From Birmingham, 47362 was on 1E72 10.30 Birmingham–Yarmouth. 47063 had worked 1E66, but the gen was that it was not coming back on 1M68, so the embarrassment of being bowled by 31260 at Thetford was avoided and Norwich was the destination. I had still never been to Yarmouth and would not be going today either as a Class 31 backed onto 1E72 at Norwich, 47362 is seen being released after the train has left for Yarmouth, while 47605 stands at the head of 1C43 15.34 to Liverpool Street.

Off to the pub it was then. Back at the station, the garishly liveried 47573 *The London Evening Standard* was at the head of what I think is 1C45 16.34 Norwich–Liverpool Street, which it will work to Ipswich. 47573 had been repainted at the start of June for the launch on Network Southeast and was named after the newspaper that was known as a stern critic of BR, in the forlorn hope of getting them onside, on 10 June. 47584 *County of Suffolk* is in the background.

Sunday 3 August 1986. The gamble of a rare Saturday night at home and a leisurely start with 47491 *Horwich Enterprise* on 1V87 10.00 Newcastle–Penzance to Birmingham paid off as 47424, rather than a no-heat one, had worked 1D54. 47484 *Isambard Kingdom Brunel* was on 1O08 13.30 Manchester–Poole to Reading, and by diagram should work 1M40 17.32 Poole–Liverpool back to Birmingham. The chance to clock up a few miles on 'the Bard' was taken; 47465 backed onto 1O08 at Reading and is seen at Basingstoke, from where 47558 May*flower* was 1M40 back. It came as no great surprise when 47558 ran round at Reading, with the Western Region taking the opportunity to keep hold of 'IKB' rather than send it back to the Midland again.

Saturday 9 August 1986. After the 47245 debacle, I had to take emergency leave upon discovering that 47242 was working 1V98 09.20 Derby–Paignton relief, and No 489 was in the book. The icing on the cake was that it made 47373 back north on 1E78 10.10 Penzance–Leeds at Exeter. Later in the day, 47529 was had to Banbury on 1O12 16.40 Liverpool–Poole for 47085 *Mammoth* to Merseyside on 1M40 17.05 Poole–Liverpool. At Lime Street, required 47296 was tantalisingly on ECS duties, while 47625 *City of Truro* was 1K07 00.15 Liverpool–Stafford, which ran via the Chester triangle.

Sunday 10 August 1986. To our horror, 47625 was removed at Crewe and an AC electric went forward as a precursor to a particularly foul overnight, even 1G00 was electrically hauled. Later in the day, 47409 *David Lloyd George* pauses at Stafford with 1M08 10.00 Portsmouth Harbour–Liverpool, which it worked throughout. 47409 had only been overhauled the previous August, it worked the Crewe Works test train on the 28th and was released on the 30th, but it turned out that it had less than one week left in traffic at this point! It later worked 1E18 18.03 Liverpool–Newcastle, before retiring to Gateshead for a B exam. It was fuelled at Gateshead in the early hours of 14 August before re-engining an overnight service to Aberdeen. That night it set off from the 'Granite City' on 1E48 21.15 Aberdeen–King's Cross, only to catch fire at Montrose, after which it never worked again.

Thursday 14 August 1986. A pre-work move saw 47468 taken to Sheffield on 1S85 07.20 Harwich–Glasgow, which dropped into the regular 1V06 11.20 Leeds–Plymouth relief. Sadly, I could only have 47357 to Derby as work beckoned.

Saturday 16 August 1986. The previous day had seen 47286 enjoyed on 1E75 14.53 Exeter–York relief from Derby to York for 47402 *Gateshead* to Bristol on 1V22 19.50 Newcastle–Newquay. Today, it was on 1Z21 09.40 Crewe–Poole relief, seen here at Wolverhampton, having crossed the Pennines overnight. 47286 had been reached off the disappointing 47646 on 1M10 06.58 Paddington–Llandudno, a train that had been getting a few no-heat locos over previous weeks. It was only a brief blast off 47286 for the magnificent 47501 to Bristol on 1V76 09.20 Liverpool–Penzance.

47234 stands at Temple Meads with 3A03 13.12 Bristol–Paddington vans. We were there for 47337 *Herbert Austin* on 1E78 10.10 Penzance–Leeds, which was taken to Sheffield for an HST to Leicester and 47100 to Birmingham on 1M76 16.00 from Yarmouth. 47113 had been on the other Norwich diagram on this day, but it would be the next day before I had a run on this one.

Sunday 17 August 1986. With 1D54 now getting '47/4s' on Sundays, it was a good chance to have a night in bed and turn attention closer to home as 1M74 07.20 Harwich Parkeston Quay–Manchester had started to get no-heat locos instead. Stratford's 47116 worked the train to Sheffield today, for Cardiff's 47241 across the Hope Valley to Piccadilly, where recently ex-works 47297 had worked the ECS for 1V13 15.00 to Swansea in from Longsight.

At the other end of 1V13, 47609 *Fire Fly*, with its unique split nameplates, was waiting to work to Cardiff, alongside 47330 on an engineers' train. Curiously, 1V13 was booked for a loco change at Cardiff, with a Class 33 diagrammed to work to Swansea.

47241 waits to work 1E87 15.20 Manchester–Harwich Parkeston Quay to Sheffield. The flush front on 47241 was the result of a collision on 9 September 1983, which put it out of traffic for over six months before re-entering service on 24 March 1984, after repairs at Crewe Works. Along with 47236, it was one of the first two Class 47s to have its boiler removed, apart from during ETH-fitment, late in 1975.

This actually turned out to be a much better Sunday move; at Sheffield, 47116 backed onto 1E87 to take it home and was taken to Ely for a DMU to March then 47113 to Leicester on 1M76 18.50 Norwich–Birmingham for an HST back to Derby. Originally built with a 12LVA24 engine, the loco had been converted to a Class 47 at Crewe Works during the first half of 1971. It had lost its trademark Stratford grey roof at its October 1984 overhaul, and it would be January 1987 before it was regained. 47116 is seen at Grantham as it waits time.

Saturday 23 August 1986. This photo caused quite a panic. 47313 had arrived early at Thetford on 1E72 10.30 Birmingham–Norwich, so the opportunity was taken to walk up to the overbridge to get a photo of it leaving, only for there to be a rumble underneath us as 47203 ran in on 1M68 12.45 Yarmouth–Birmingham extremely early! Passenger trains are not supposed to leave early, but at an outpost like this, anything was possible. Cue a mad dash back to the station, stopping near where 47313 was parked to pick up the bag on the way. And, yes, we did make it!

Sunday 24 August 1986. The 1M74 07.20 Harwich Parkeston Quay–Manchester had got its booked Tinsley Class 47/4 with 47439 working to Sheffield. It was actually booked for a no-heat Western Region '47' across the Hope Valley, but today Bescot's 47002 was provided and is seen here backing onto the stock in readiness to work 1E87 back to Sheffield.

The ECS for 1V13 15.00 Manchester–Swansea had once again been worked in from Longsight by a no-heat '47'. 47050 had been the first of the class to be painted into Railfreight livery at Crewe Works and, as such, was slightly non-standard with large hand-painted numerals and the windows picked out in black.

Cardiff Canton-allocated 47559 *Sir Joshua Reynolds* prepares to head for home on 1V13 15.00 Manchester–Swansea, which it will work to Cardiff. It had started the previous day in Wales, but in the north of the country, working 1A52 10.10 Holyhead–Euston to Crewe. The next day would see it working to the English capital on a 1Z65 15.28 Exeter–Paddington relief. It would finish Tuesday night on the south coast off 1O12 16.40 Liverpool–Poole.

47439 worked 1E87 forward from Sheffield to Harwich and is seen here during its Nottingham stop. The next day saw it work 1S85 07.20 Harwich–Glasgow to Sheffield, before returning to the Suffolk port on 1E87 10.20 Glasgow–Harwich. It set off on 1S85 again on the 26th, but it failed at Ipswich with a power earth fault and the train was worked forward by 47634 *Henry Ford*. The fault, however, was soon rectified and it was on 1H06 08.35 Liverpool Street–King's Lynn on the 27th.

Taking 1E87 to March allowed time for a stroll to the nearby depot, where there was not a soul about, so the opportunity was taken for a wander round to get a few photos. 47324 is seen at the head of a row of locos, ahead of 47060 and 47378. 47324 sports a flush front at its No 1 end, a legacy of a crash on 12 April 1985, when a brake failure led to it running away at Oxley while it was stabled on an MGR train. It was rebuilt at Crewe Works over the next four months.

47256 had been the last Class 47 in green, albeit block green rather than two-tone. It was repainted into block green at Cardiff Canton in December 1977 following fire damage and was finally painted blue at its October 1978 overhaul, almost 12 years after D1953 had been released from Brush in blue. 47256 was overhauled at Doncaster in June 1987 when it remained blue rather than going into Railfreight livery, as it was a Departmental sector loco at the time. It had been on passenger duty the previous Monday when it worked 2G24 07.40 Cardenden–Edinburgh prior to working south on 4M77 1420 Bathgate–Washwood Heath empty 'Cartics'.

47378 had been the third of the class to be released from Crewe Works in Railfreight livery in September 1985, following 47050 and 47095. It still had the large painted numerals, but the black around the windows had been modified.

Tuesday 26 August 1986. A foul August day, which was almost as foul as my mood as I got to Derby just in time to see 47340 leaving on a 1V40 08.50 Derby–Paignton relief. I followed it to Birmingham on 1V41 07.00 Bradford–Paignton HST in order to get to 47147 on 1E51 08.56 Bristol–Sheffield relief, seen here having terminated in Platform 8. Note the basher's bag on the trolley and the gaggle of bashers taking refuge from the rain under the canopy. The flush front stems, it is believed, from an accident sustained almost as soon as it left works following overhaul on 20 November 1982, as it was back in two days later for unclassified attention. It was released again on 18 December.

After running round, 47147 waits to return to the Western Region with 1V18 13.08 relief to Cardiff, which dropped me nicely into a 14.00 backshift in the Travel Centre. The pouring rain can be seen in the photo.

Inverness's 47614 at Derby on 1E78 16.52 Birmingham–Leeds. On the night of the 24 August, 47614 had set off on 1E42 23.40 Edinburgh–King's Cross, but failed at Newcastle with the blue fault light on the driver's desk shining bright, a sign of high engine coolant temperature, blower motor failure, loss of air or vacuum pressure, or an overload relay tripped. 47403 worked 1E42 forward and 47614 went to Gateshead for repair before later working 1M76 14.19 Newcastle–Liverpool and 1E89 20.44 Liverpool–York. It had started this morning on 2P06 08.35 York–Scarborough, then 1M67 09.53 Scarborough–Liverpool, and 1G61 14.30 Liverpool–Birmingham. 47614 would be back at Gateshead for investigation into low water pressure early on the 28th after arriving on 3E07 21.55 Bristol–Newcastle. The Eastern Region then repatriated it on that night's 1S37 19.01 King's Cross–Aberdeen vans, which it worked from Newcastle.

Wednesday 27 August 1986. 47634 *Henry Ford* was on its second day on 1S85 07.20 Harwich Parkeston Quay–Glasgow, following the failure of 47439 the previous day. The '47' on the train had worked through to Preston in the past, but in the 1986 timetable, it was booked for an engine change at Sheffield to avoid the run round, with the loco that worked in returning to Suffolk on 1E87 10.20 Glasgow–Harwich. Ex-works in early July, 47518 has backed onto 1S85 to work to Preston. 47652 on 1V06 11.20 Leeds–Plymouth relief would take me to my backshift at work.

Friday 29 August 1986. Putting my 14 required Class 47s through TOPS at lunchtime on a morning shift put the cat among the pigeons as 47097 was allocated to 1C73 19.07 Paddington–Plymouth. The sensible thing would have been to take an HST to St Pancras to watch the commuter trains out of Paddington but, not fancying an HST to London, I opted for the 15.38 HST to Birmingham for 47558 May*flower* on 1O46 15.10 Liverpool–Portsmouth Harbour to Reading, hoping that 47097 did not work an earlier train. My luck was in and 47097 worked 1C73 and is seen at Plymouth some 187 miles later. Return north was with 47520 on 1E58 00.05 Plymouth–York.

Saturday 30 August 1986. It was a curious day on the Yarmouths with 47227 on 1E66 08.19 off New Street, but 47592 *County of Avon* on 1E72 10.30 departure. The former is seen at Nuneaton; no one used to bat an eyelid at bashers going off the end of the platform for a better photo, but it would be a different case nowadays. 47534 on 1V45 08.29 Nottingham–Paignton was the move back to Birmingham.

In the meantime, 47369 had worked 1O05 06.52 Leeds–Poole out of Birmingham. Fortunately, 1O08 09.12 York–Poole was also no-heat out of Birmingham, with Kingmoor's 47150 working it to Reading before going forward on 1O09 08.12 Newcastle–Weymouth. 47150, seen here at Oxford, had transferred away from Stratford in March 1984 but still bore the trademark of its former home. Its flush front was a souvenir of when its driver had backed onto 6M38 02.15 Port Clarence–Long Eaton too enthusiastically on 6 December 1985; it had only been back in traffic since 1 July following repairs at Crewe.

47369 blasts out of Crewe on the last leg of 1M14 11.40 Poole–Liverpool. It had also been in passenger action the previous day, working 1M88 06.25 Poole–Manchester to Reading before taking over from 47501 on 1S39 08.34 Poole–Glasgow and working to Birmingham. The day was rounded off with 47638 to Banbury on 1O12 16.40 Liverpool–Poole and 47462 back to Birmingham on 1M40 17.05 Poole–Liverpool.

47326 runs into Crewe with a UKF fertiliser train from Ince & Elton. As D1807, the loco had been the first Class 47/3 to be delivered to the London Midland Region, and the first with Serck radiator louvres when it was delivered from Brush on 30 January 1965. 47326 would be due for overhaul in 1987 but the work was denied, and the loco was stopped with engine defects on 3 February 1987. It was moved to Bristol Bath Road for storage with its future looking bleak, but it was to be saved by the demise of Frome accident victim 47202, which donated its recently overhauled power unit to 47326 in September 1987.

Sunday 31 August 1986. What had been a very good weekend turned into a great one with the gen that 47245 was allocated to 1O46 14.20 Liverpool–Portsmouth Harbour. 47618 *Fair Rosamund* was 1M74 07.20 Harwich–Manchester to Sheffield for 47375 across the Hope Valley. 47652 on 1V13 15.00 Manchester–Swansea dropped into 1O46 nicely at Crewe and No 493 was in the book to Banbury, tasting all the sweeter for having missed it a few weeks earlier. The weekend was rounded off with 47423 back to Birmingham on 1M25 16.00 Portsmouth Harbour–Manchester.

Saturday 6 September 1986. After an overnight involving 47544 to Bristol on 1V22 19.50 Newcastle–Newquay for 47599 back on 1E58 00.05 Plymouth–York, I found myself briefly heading south with 47448 on 1O01 06.26 Wolverhampton–Poole. I really should have carried on south but returned to Birmingham for 47337 *Herbert Austin* to Norwich on 1E66, and it is seen here at Norwich, waiting to return to Birmingham on 1M68 12.45 from Yarmouth. 1E72 failed to produce again, with a pair of Class 31s on the train. The reason I should have continued south? One of my winners, 47094, worked 1B42 12.10 Paddington–Pembroke Dock, which I knew nothing about at the time.

Wednesday 10 September 1986. A rose between two thorns, with a third thorn in the shed just for good measure. With freshly greased buffers, a faded looking 47328 rests at Toton. The Class 45s certainly were a thorn in my side, but, from the following May, they would no longer have weekday passenger diagrams; curiously, in a way, this made bashing less fun with much of the jeopardy removed. 47328 had been a LMR engine from new but would move to Stratford in October 1987, followed by transfer to Eastfield in August 1990. It would remain without a headlight until April 1987.

Saturday 13 September 1986. Fortunately, the station buffet at Bristol used to be open all night Friday into Saturday during the summer. I arrived there around midnight with 47625 *City of Truro* on 1C82 19.30 Milford Haven–Bristol, and the buffet provided a place of refuge with a couple of hours to kill prior to heading north with 47541 *The Queen Mother* on 1E58 00.05 Plymouth–York, seen at its destination. The Inverness machine had been released off repairs at Holbeck the previous morning and had run light to York to work 1V76 13.18 to Plymouth. It would return west later, working 1V61 10.05 York–Pembroke Dock to Gloucester.

This was to be my lucky day. 47351 appeared vice an HST on 1S02 07.30 Leeds–Aberdeen, and I was planning on a high mileage '47/3' day with a trip to the 'Granite City'. However, the guard came round out of York saying that anyone travelling north of Newcastle would need to change there, as the train was being swapped out for an HST. A quick change of plans and I was off at Darlington for 47521 to Peterborough on 1P38 08.43 Newcastle–Yarmouth, 'blind' (as in with no gen) for 1E72 10.30 Birmingham–Yarmouth. A major stroke of luck saw required 47291 on 1E72, seen here departing Thetford. 47291 had transferred to Stratford the previous September and the worry was that it would not come back from Norwich on 1M76. No 494 in the book and the last one for 1986.

47226 was the power back from Thetford on 1M68 12.45 Yarmouth–Birmingham, seen during a crew change at March. It would return east on 1E82 18.25 Birmingham–Norwich, which connected into 1S70 20.12 King's Cross–Aberdeen, powered by 47422 to Newcastle. Here the train reversed and continued to Scotland behind 47618 *Fair Rosamund*, which found itself far from home at Aberdeen in the morning.

Sunday 14 September 1986. The best thing about Dundee was always leaving the place, even better in the front compartment of a de-classified coach, right behind a steaming '47' on a crisp autumnal day. 47209 simmers nicely as it waits to depart on 2J56 08.30 Dundee–Edinburgh. It was a good day for steam heat on 'the circuit', with 47206 on 2J53 09.20 Edinburgh–Dundee and 47152 on 2J61 13.20 off Edinburgh. 27049 was the 11.20 ex-Edinburgh, but the task proved too much, and it was piloted by 26004 on the 15.30 return.

47152 has arrived at Edinburgh with 2J62 16.30 ex-Dundee. The plan was to take the 18.30 to Glasgow for the 19.25 Glasgow–Aberdeen to Dundee for 1E43 20.00 Aberdeen–King's Cross, but this fell apart as 47713 *Tayside Region* was halted before Linlithgow due to the failure there of 47714 *Grampian Region* on the 18.00 ex-Edinburgh. Unusually, 47713 was leading and, with 47714 at the Edinburgh end, made for an incongruous sight as the two trains were coupled together with the locos in the middle. After a lot of shunting about, arrival into Queen Street was at 19.58. With the 19.25 to Aberdeen long gone, departure was at 20.03 with 47706 *Strathclyde* on the slightly delayed 20.00 to Edinburgh.

Monday 15 September 1986. 1E43 was 47544 to Newcastle for 47566 on to London. 47566 had been in trouble on Saturday when it failed at Aycliffe on 1E99 11.12 Bangor–Newcastle, and had been rescued by a pair of Class 37s, but it was back on fine form, arriving into 'the Cross' at 06.22, some 48 minutes ahead of its booked time! 47482 is seen with the ECS off 1A33, the previous night's 23.00 Newcastle–King's Cross.

1H06 08.35 Liverpool Street–King's Lynn was booked for a Gateshead Class 47 rather than a Stratford one, off the previous night's 4C64 Follingsby–Stratford Freightliner or, in this case, as it was a Monday, off Friday's 4C64. After a weekend on Cambridge services, 47651 pauses at the university town on 1H06 with a Class 105 DMU on the 10.00 to Ipswich in the bay. If it sticks to diagram, 47651 should work 1S79 22.15 King's Cross–Aberdeen out of London on Tuesday night.

Former Stratford charge, 47457 *Ben Line*, had been one of those made redundant in East Anglia in May and had moved to Gateshead. It had, however, made a return today and waits in the rain to work 1H30 17.35 Liverpool Street–Cambridge. 47465 of Crewe was another foreigner on the route that day, along with locals 47573, 47577, 47582 and 47649.

Tuesday 16 September 1986. Gateshead's 47568 and Bescot's 47337 *Herbert Austin* rest between duties at Aberdeen. 47568 would later work 1E43 to Newcastle for 47432 forward. 47049 was the main attraction on this day, its boiler put to good use while working 1A46 08.25 Aberdeen–Inverness and 1H29 11.35 return.

Wednesday 17 September 1986. 47572 had been named *Ely Cathedral* at Ely on 16 May 1986, but here it was at Liverpool Street, seemingly involved in some sort of dedication ceremony in between working in on 1C79 09.00 from King's Lynn, and returning from whence it came on 1H16 at 12.35. 1H06 had been 47654 on this day with 47487, 47524, 47573 and 47577 also in action on the Cambridge line. This was a dark day personally as 47282 was officially withdrawn, leaving me with a maximum number of 505 Class 47s for haulage.

Thursday 18 September 1986. The journey to Scotland on 1S79 22.15 King's Cross–Aberdeen was unusual as 47555 *The Commonwealth Spirit* worked through to Edinburgh, rather than being swapped at Newcastle, as Gateshead had run out of Class 47s. Carrying on with 1S79 to Dundee, 47644 *The Permanent Way Institution* worked forward. 47562 *Sir William Burrell* was on 2J48 07.20 Dundee–Edinburgh, along with another ETH machine, 47426, on 2J25 08.15 Edinburgh–Dundee. Steam heat was eventually found in the shape of 47209 on 2J08 08.30 Dundee–Edinburgh. Back at 'Auld Reekie', 47004 was lined up for 1L35 11.23 Edinburgh–Perth, which at least gave a break from the monotony of 'the circuit', and is seen at the 'Fair City', having run round ready to work 1T24 13.50 Perth–Glasgow.

It always seemed to be impossible to go to Scotland without bumping into 47481 at some point, despite it being an LMR machine. It had, however, moved closer to Scotland now, following its transfer from Crewe to Carlisle Kingmoor. Here it is at Perth on 1C84 11.00 Inverness–Carstairs, the portion for 'The Clansman', which now also conveyed coaches from Glasgow. This was, of course, 'flagged' for 47004 to Glasgow, then to Dundee on 1L37 16.03 Glasgow–Arbroath. 47205 on 2J24 18.30 Dundee–Edinburgh, and 47012 on 2J33 20.15 Edinburgh–Dundee to Leuchars for the 'Nightrider' to London rounded off a good steam heated day.

Saturday 20 September 1986. After a quick blast to Wolverhampton with 47285 on 1J18, the main event of the day was to be 47005 to Norwich and back on 1E66/1M68. However, at Norwich, Stratford's required 47054 was on the blocks with the 10.30 ex-London. Going into the TOPS office to enquire as to what 47054 was doing next, it turned out that 47580 had failed at Ipswich and that 47054 would work 1C37 14.34 Liverpool Street to Ipswich. 47005 was allowed to leave at 13.30 and sure enough 47054 backed onto 1C37. At about 14.30, the guard started kicking off about 47054 being unable to power the air conditioning and that he wanted a fresh engine. The station supervisor told him that there were no other locos available and that he would have to take it, the air conditioning would be working beyond Ipswich after all. It seemed that we were about to depart when 47605, which had failed earlier in the day with a ruptured oil pipe, rolled into the station yard and the guard kicked off again, 'I want that one'. To cut a long story short, the guard got his way this time and the shunter got in between 47054 and the front coach to uncouple and the loco swiftly disappeared. 47054 would have been my 495th Class 47, so well and truly bowled! With no other options left, I took 47605 to Ipswich for 47617 *University of Stirling* back on 1P48 14.30 Liverpool Street–Norwich off 47605. 47218 was on 1M76 16.00 Yarmouth–Birmingham forward and this was had to Melton Mowbray for 47005 back to Peterborough on 1E82. It turned out that there had been some excitement on 1M68 earlier as the driver of 47005 had taken the 40mph curve at Wigston at excessive speed, to the extent that someone had pulled the communication chord as they feared a derailment. The tin lid was put on my day when I was stood at Peterborough with 'Bicester' and 47558 May*flower* rolled in bang on time on 1S70, only for the shunter to appear once again to remove it, with 47412 being its replacement. 'Bicester' was delighted, me less so. After several overnights, a cut price sleeper berth was negotiated with the steward and at least a good night's sleep was had to Aberdeen. 47412 gave way to 47568 at Newcastle in the early hours with the train reversing to go via Carlisle again.

Sunday 21 September 1986. 47643 was had on 1T78 10.50 Aberdeen–Glasgow for 47460 on 2J52 12.30 Dundee–Edinburgh. After 47562 on 1A89 14.55 Edinburgh–Aberdeen, steam was found atop 2J58 15.30 Dundee–Edinburgh in the shape of 47209. A leaky steam pipe gives the impression that 2J57 17.20 Edinburgh–Dundee was only load two as 47004 waits time at Inverkeithing. From Dundee, 47710 *Sir Walter Scott* was 1A91 17.25 Glasgow–Aberdeen to Stonehaven for 47568 on 1E43 20.00 Aberdeen–King's Cross to Newcastle.

Monday 22 September 1986. 47419 re-engined 1S79 at Newcastle and was my ride back to Edinburgh, where I took a gamble on 47205 on 2G01 06.05 to Cowdenbeath. Not fancying Cowdenbeath at 06.52 in the morning, I managed to secure a ride on the ECS to Thornton Junction and back to Cardenden, where 47205 waits time before working 2G24 07.40 to Edinburgh.

Another day on 'the circuit' ensued. Two round trips were managed with 47004 on the 10.15 and 16.15 ex-Edinburgh. 47579 *James Nightall GC* was a stranger on the route and is seen leaving Markinch on 2J13 14.15 Edinburgh–Dundee, and 47586 was coming back south on 2J14 14.30 Dundee–Edinburgh. The latter was then swapped out at Edinburgh and 47209 worked the next leg of its diagram, 2J31 18.15 ex-Edinburgh.

Tuesday 23 September 1986. 1H06 actually got a Stratford Class 47 today rather than the booked Gateshead one. The railman, without a high visibility vest but with a hat, operates the groundframe at King's Lynn in order that 47576 can run round the stock to form 1C81 11.05 back to Liverpool Street. This was a black day as 47464 failed east of Elgin while working 1H27 09.35 Aberdeen–Inverness and was rammed by 37416, which had been sent to rescue it. 47464 was deemed beyond economic repair and became the first Class 47/4 withdrawal outside of the original 47401–420 batch.

Friday 26 September 1986. 1M73 11.24 Newcastle–Liverpool was the best bet for a Class 47 on a Pennine turn at Newcastle, as it was part of an ECML diagram. The loco was booked off 1N16 00.05 King's Cross–Newcastle, after being serviced at Gateshead. It was booked to return on 1E59 18.03 Liverpool–Newcastle and, after fuelling at Gateshead again, was booked to work 1S37 19.01 King's Cross–Aberdeen forward at 01.12. The problem was that the London Midland Region had a tendency to send a Class 45 back on 1E59 and keep the Class 47 that had worked in. This was the subject of several complaints from York to Crewe as loco changes could not be affected at Newcastle due to a shortage of '47s'. 47412 stands in Platform 10 on 1M73 while 47652 is in 'the hole' on 3N04 Sunderland vans. 47412 did return on 1E59 but did not work 1S37, as it was stopped at Gateshead with low power.

Saturday 27 September 1986. As an Immingham engine, 47222 *Appleby-Frodingham* was always a treat on passenger trains. 47222 had started the day on 1E61 09.40 Poole–Bradford Interchange to Reading, where 47441 took over. It then worked 1E63 09.55 Weymouth–Newcastle to Birmingham, which it took over from 47426.

Appendix

The Birmingham Area Booked NB Turns 1986 (all booked Bescot Class 47)

Date	1J18	1E66 - 1M68 - 1E82	1E72 - 1M76
17/5/86	37426	47346	47085
24/5/86	47209	47115	47560
31/5/86	47204	47346	47361
7/6/86	47157	47195	47131
14/6/86	47234	47200 (note 1)	47249
21/6/86	47063	47198	47482
28/6/86	47532	47007	47594
5/7/86	47632	47322 (note 2)	47318
12/7/86	47220	47350	31411+31419
19/7/86	47369	47162	47113
26/7/86	47231	31105+31162	47357 (note 4)
2/8/86	45041	47063 (note 3)	47362
9/8/86	47338	47362	47626
16/8/86	37242	47113	47100
23/8/86	37244	47203	47313 (note 5)
30/8/86	37241	47227	47592
6/9/86	31155	47337	31181+31288
13/9/86	37069	47226	47291
20/9/86	47285	47005	47218
27/96	47115	47315	47590

Workings

1J18 07.25 Birmingham–Aberystwyth to Shrewsbury.

1E66 08.19 Birmingham–Yarmouth to Norwich.
1M68 12.45 Yarmouth–Birmingham from Norwich (depart 13.30).
1E82 18.25 Birmingham–Norwich.

1E72 10.30 Birmingham–Yarmouth to Norwich.
1M76 16.00 Yarmouth–Birmingham from Norwich (depart 16.40).

Notes

1) 31285 worked 1M68 & 1E82
2) 47577 worked 1M68 & 1E82
3) 31260 worked 1M68 & 1E82
4) Failed at Syston Junction, 31275 forward, 31308 worked 1M76
5) 31219 worked 1M76

1E07 07.10 Sheffield–Skegness and 1E02 10.43 return were nominally booked for a no-heat Crewe Class 47, but 47305 on 7 June (page 53) and 47358 two weeks later were the only ones to work the train. 47476 on 5 July was the only other Class 47 to work the diagram. The only other Saturday NB Class 47 diagram was a dated service between Preston and Blackpool.